People and Forests

Politics, Science, and the Environment
Peter M. Haas, Sheila Jasanoff, and Gene Rochlin, editors

Shadows in the Forest: Japan and the Politics of Timber in Southeast Asia, Peter Dauvergne

Views from the Alps: Regional Perspectives on Climate Change, Peter Cebon, Urs Dahinden, Huw Davies, Dieter M. Imboden, and Carlo C. Jaeger, editors

People and Forests: Communities, Institutions, and Governance, Clark C. Gibson, Margaret A. McKean, and Elinor Ostrom, editors

People and Forests

Communities, Institutions, and Governance

edited by
Clark C. Gibson, Margaret A. McKean,
and Elinor Ostrom

The MIT Press
Cambridge, Massachusetts
London, England

Chapter 5, "Optimal Foraging, Institutions, and Forest Change: A Case from Nepal," by Charles M. Schweik, is reprinted with permission from Kluwer Academic Publishers, forthcoming in *Environmental Monitoring & Assessment* 63/64 (2000).

This book was set in Sabon by Achorn Graphic Services, Inc., and printed and bound in the United States of America.

Printed on recycled paper.

Library of Congress Cataloging-in-Publication Data

People and forests : communities, institutions, and governance / edited by Clark C. Gibson, Margaret A. McKean, and Elinor Ostrom.
    p.   cm.—(Politics, science, and the environment)
  Includes bibliographical references and index.
  ISBN 0-262-07201-7 (hc.: alk. paper)—ISBN 0-262-57137-4 (pbk.: alk. paper)
  1. Forest management—Social aspects. 2. Forestry and community.
I. Gibson, Clark C., 1961–   II. McKean, Margaret A. III. Ostrom, Elinor.
IV. Series.

SD387.S55 P46   2000
333.75—dc21
                                                          99-043446

To Marilyn Hoskins, who inspired our work by her dedication to improving the lives of villagers living in and near forests while encouraging rigorous policy-related research, and to the members of the communities around the world who made our work possible by welcoming us into their homes and forests.

Contents

Tables

Figures

Series Foreword

As our understanding of environmental threats deepens and broadens, it is increasingly clear that many environmental issues cannot be simply understood, analyzed, or acted on. The multifaceted relationships between human beings, social and political institutions, and the physical environment in which they are situated extend across disciplinary as well as geopolitical confines and cannot be analyzed or resolved in isolation.

The purpose of this series is to address the increasingly complex questions of how societies come to understand, confront, and cope with both the sources and manifestations of present and potential environmental threats. Works in the series may focus on matters political, scientific, technical, social, or economic. What they share is attention to the intertwined roles of politics, science, and technology in the recognition, framing, analysis, and management of environmentally related contemporary issues, and a manifest relevance to the increasingly difficult problems of identifying and forging environmentally sound public policy.

Peter M. Haas
Sheila Jasanoff
Gene Rochlin

Preface

The road we took to write this book has been somewhat long and involved. It actually started with a long-term study of irrigation institutions by the Workshop in Political Theory and Policy Analysis. Marilyn Hoskins, former head of the Forests, Trees, and People Programme at the United Nations Food and Agriculture Organization (FAO), had heard about the Workshop's database on irrigation institutions. This database allowed us to understand how various kinds of governance arrangements affected the performance of irrigation systems (see Shui-Yan Tang, *Institutions and Collective Action: Self-Governance in Irrigation*, San Francisco: ICS Press, 1992; Wai Fung Lam, *Governing Irrigation Systems in Nepal: Institutions, Infrastructure, and Collective Action*, Oakland, CA: ICS Press, 1998). She inquired whether we might be willing to undertake a similar effort to study various types of forests that were governed by local communities, by national and regional governments, and by private individuals. In 1992, she sent a group of scholars including James Thomson, Gabriel Campbell, Arun Agrawal, Margaret McKean, Rajendra Shrestha, and Jean-Marc Boffa to Bloomington to explore the possibility of establishing a research program to study forest resources and institutions throughout the world. Mary Beth Wertime, George Varughese, Paul Turner, Sharon Huckfeldt, and Elinor Ostrom represented the Bloomington team as we began to think through the implications of doing systematic research in the complex world of forests. Other valuable participants were Paul Benjamin, Ganesh Shivakoti, Minoti Chakravarty-Kaul, David Green, and Vincent Ostrom. The challenge was both engaging and daunting. Our fascination with the questions that we have been addressing has not wavered even though the challenges have, at times, been almost overwhelming.

It took several years just to design and pretest the set of instruments that are at the core of the International Forestry Resources and Institutions (IFRI) method. We were fortunate that it was possible for Rosario León to pretest these instruments in Bolivia, for Rajendra Shrestha to pretest them in Nepal, for Hamidou Magassa to pretest them in Mali, and for Arun Agrawal to pretest them in India. During this period of design and pretest, we received comments from over 40 scholars from multiple disciplines representing all regions of the world who reviewed various drafts of the research instruments. Subsequently, the first pilot studies were conducted in India and Bhutan by Arun Agrawal and in Uganda by William Gombya-Ssembajjwe, Abwoli Banana, Dusty Becker, David Green, and Elinor Ostrom during the fall of 1993.

Our initial image of what we conceptualized the IFRI research program to be was written by Elinor Ostrom and Mary Beth Wertime in 1994 as "The IFRI Research Strategy." We shared this vision with prospective Collaborating Research Centers as we slowly built an effective research network. Research proposals to the Forests, Trees, and People Programme at FAO in Rome, to the Ford Foundation, and to the MacArthur Foundation owe much of their substance to the research strategy. We have reproduced that document as the appendix to this book to share with others what we were thinking about as we designed the overall research program.

Sharon Huckfeldt designed the IFRI relational database, with substantial input from Charlie Schweik and Julie England along with the help of Tom Koontz, Paul Turner, George Varughese, and other hard-working graduate assistants who faithfully tried to think of every possible test they could to challenge the system while it was in design phase. The result of all of their hard work has been a remarkably robust relational database that has successfully survived torturous field conditions. Since then, the responsibility for the database has been shouldered ably in turn by Joby Jerrells, Julie England, and Robin Humphrey, respectively. Robin continues to be responsible for the day-to-day management of the database and ensures the reliability of the database on which we all depend. While Robin and Julie (now as systems analyst) have many responsibilities, their dedication to the database has made it possible to continue to stay on top of what would otherwise have become an impossible data-management task.

We began to hold an annual training program for scholars interested in the IFRI research program in 1994. This program is now offered as a regular Indiana University two-month training program each September and October, when visiting scholars and local Ph.D. students conduct a joint study of local forest institutions as they learn how to apply the concepts that underlie the structure of the IFRI database. In addition, we have tried to bring together the members of the IFRI research network once a year depending on obtaining financial support for this essential activity. The first meeting was hosted by Mike Arnold at the Oxford Forestry Institute in December of 1994. The second meeting was held in conjunction with the International Association for the Study of Common Property (IASCP) meetings in the Philippines in 1995. The third meeting was held in conjunction with the IASCP meetings in Berkeley in 1996; the fourth meeting was held in conjunction with the IASCP meetings in Vancouver in 1998; and the fifth meeting is to be held in conjunction with the Second Workshop on the Workshop conference in Bloomington in June of 1999. During many of these meetings we have received sage advice from Mike Arnold, Gabriel Campbell, Marilyn Hoskins, and Narpat Jodha.

The contact persons and the affiliated institutions of the IFRI research network are:

Bolivia
Rosario León
Centro de Estudios de la Realidad Economica y Social (CERES)
Cochabamba, Bolivia

Kenya
Paul Ongugo
Kenya Forestry Research Institution (KEFRI)
Nairobi, Kenya

Madagascar
Roland Raharison
Universiti di Antananarivo
Antananarivo, Madagascar

Mali
Robert Dembele
Centre Technique d'Execution des Programmes d'Action (CTEPA)
Bamako, Mali

Nepal
Mukunda Karmacharya and Birendra Karna
Nepal Forestry Resources and Institutions (NFRI)
Kathmandu, Nepal

Tanzania
George Kajembe
Department of Forest Mensuration and Management
Sokoine University
Morogoro, Tanzania

Uganda
William Gombya-Ssembajjwe and Abwoli Banana
Uganda Forestry Resources and Institutions Center (UFRIC)
Department of Forestry
Makerere University
Kampala, Uganda

The Workshop in Political Theory and Policy Analysis at Indiana University initiated and coordinated the IFRI research program. This role changed somewhat when the National Science Foundation announced a special centers competition related to the study of human dimensions of global change in 1995. Elinor Ostrom of the Workshop and Emilio Moran of the Anthropological Center for Training and Research on Global Environmental Change combined efforts with two other major research centers on the Bloomington campus to propose a Center for the Study of Institutions, Population, and Environmental Change (CIPEC). After a vigorous national competition, CIPEC received funding for an initial five years in May of 1996. Since then, CIPEC has used the IFRI research instruments as one of the core set of measurements in all of its major sites in the Western Hemisphere. The interpretation of remotely sensed images and the use of geographic information systems are additional tools used in conjunction with the IFRI research instruments in all major CIPEC sites. Thus, the IFRI research program is now cosponsored by both the Workshop and CIPEC at Indiana University for research conducted in both hemispheres.

Throughout this adventure, we have been dependent on two sources of support. The first has been the large number of forest users who have welcomed us into their homes, spent hours with us in the field, helped us in pacing off forest plots, helped identify trees and other vegetation, made sure we found our way to and from forests, provided us with housing and food, and commented on the draft reports that we have sent to them from our individual studies. We have also had very high levels of cooperation from forest officials at all levels and local researchers in the countries where members of the IFRI research network have conducted research. We are hopeful that the reports we have provided back to local communities and to government offices have been of sufficient value that they will welcome us back when we return to revisit these sites as part of our long-term research strategy.

The second source of support has come from funding institutions that have provided us resources throughout the life of the IFRI research program. The entire program would have been inconceivable but for their support throughout the years. We wish particularly to thank Marilyn Hoskins, Krister Andersson, Hivy Ortiz, Katherine Warner, and Alice Ennals at FAO in Rome, who have supported this research program, as well as Carlos Brennes of the Forests, Trees, and People Programme in Costa Rica and Gustavo Gordillo de Anda, assistant director-general, FAO regional representative for Latin America and the Caribbean. We also thank the Ford Foundation, the MacArthur Foundation, and the National Science Foundation for their support of individual projects within this broad research program and the National Science Foundation for its support of CIPEC. We also want to thank our many colleagues throughout the world and the graduate and undergraduate students at Indiana University for the hard work and thoughtful criticisms they have provided to the IFRI research program over the years. Special thanks go to Patty Dalecki for her invaluable skills and efforts in the editing and preparation of this volume.

Clark Gibson, *Bloomington, IN*

Margaret A. McKean, *Durham, NC*

Elinor Ostrom, *Bloomington, IN*

Contributors

Arun Agrawal teaches comparative politics at Yale University. He has done empirical research on forest-using communities and pastoralists in India and Nepal and has also written on narrative formations around development, environmental conservation, and community and indigenous knowledge and on the theoretical aspects of institutions and common property. His book *Greener Pastures: Politics, Markets, and Community among a Migrant Pastoral People* (1999) focuses on nomadic shepherds in India.

Abwoli Y. Banana is a senior lecturer in the Department of Forestry, Makerere University, and a research associate at the Makerere Institute of Social Research (MISR). He is coleader of the Uganda Forestry Resources and Institutions Center, which is studying the relationship between local communities and their forests in Uganda.

C. Dustin Becker is a professor at the Department of Horticulture, Forestry, and Recreation Resources at Kansas State University. She is continuing her work regarding local-level management of natural resources in Ecuador by, among other things, studying land tenure issues among local-level farmers and examining the moisture-trapping characteristics of coastal forests.

Clark C. Gibson is an assistant professor of political science at Indiana University and a research associate with the university's Center for the Study of Institutions, Population, and Environmental Change. A large portion of his research explores the institutions and politics of natural resources, especially forest and wildlife resources in Africa and Latin America, at both local and national levels. His recent book *Politicians and Poachers: The Political Economy of Wildlife Policy in Africa* (1999) examines the politics of wildlife management at multiple levels in Zambia, Kenya, and Zimbabwe.

William Gombya-Ssembajjwe is the leader of the Uganda Forestry Resources and Institutions Center at Makerere University and a senior lecturer in the Forestry Department there.

Rosario León is a sociologist at the Centro de Estudios de la Realidad Economica y Social (CERES) in Cochabama, Bolivia. As the facilitator of the FTPP in Bolivia, she has been working with IFRI since 1994.

Margaret A. McKean is a member of the Department of Political Science and the Nicholas School of the Environment at Duke University, where she teaches environmental politics and policy and Japanese politics. She works on the evolution of common property rights and on politics as a struggle over the production of collective goods, in Japan and elsewhere. She is the author of *Environmental Protest and Citizen Politics in Japan* (1981) and coeditor of *Making the Commons Work* (1992).

Elinor Ostrom is codirector of the Workshop in Political Theory and Policy Analysis, the Center for the Study of Institutions, Population, and Environmental Change (CIPEC) and the Arthur F. Bentley Professor of Political Science at Indiana University. She is the author of *Governing the Commons* (1990) and *Crafting Institutions for Self-Governing Irrigation Systems* (1992), coauthor with Larry Schroeder and Susan Wynne of *Institutional Incentives and Sustainable Development* (1993), and coauthor with Roy Gardner and James Walker of *Rules, Games, and Common-Pool Resources* (1994).

Charles M. Schweik is an assistant research professor with the Center for Public Policy and Administration and the Resource Economics Department at the University of Massachusetts, Amherst. His interest areas are in information technology, environmental policy, public management, and research methodology. His recent research focuses on the human dimensions of environmental change, specifically applying geographic information systems (GIS) and satellite imagery analysis to study human incentives, actions, and environmental outcomes. He is also studying forest and irrigation management, both in the United States and Nepal.

George Varughese is program development advisor for the United Nations Development Programme in Nepal. He has recently researched the organization of collective action for the governance and management of forest resources in Nepal from an institutional analysis perspective. He is also interested in the institutional design of partnerships between local communities and government officials for local governance, participatory management of natural resources, and the delivery of social services. Most recently, he has coedited the book *People and Participation in Sustainable Development: Understanding the Dynamics of Natural Resource Systems* (1997) with Ganesh Shivakoti, Elinor Ostrom, Ashutosh Shukla, and Ganesh Thapa.

Mary Beth Wertime served as the International Forestry Resources and Institutions (IFRI) research program coordinator and as a research associate at the Workshop in Political Theory and Policy Analysis until 1996. She has worked in Cameroon and Mali on agroforestry program planning and evaluation and in Ghana, the Ivory Coast, and countries in Latin America over the past 15 years.

Acronyms and Abbreviations

CBS	Central Bureau of Statistics
CENR	Committee on Environment and Natural Resources
CERES	Centro de Estudios de la Realidad Economica y Social
CGIAR	Consultative Group on International Agricultural Research
CIPEC	Center for the Study of Institutions, Population, and Environmental Change
CRC	collaborating research center
DBH	diameter at breast height
DFO	district forest office
DGPS	differential global positioning system
FTPP	Forests, Trees, and People Programme
GIS	geographic information system
GPS	global positioning system
IAAS	Institute of Agriculture and Animal Science
IAD	institutional analysis and development
IFRI	International Forestry Resources and Institutions
IRR	incident-rate ratio
MLE	maximum likelihood estimation
NAS	National Academy of Sciences
NEAP	national environmental action plan
NGO	nongovernmental organization
NSF	National Science Foundation
OLS	ordinary least squares

PRA	participatory rural appraisal
RC	residence council
UFRIC	Uganda Forestry Resources and Institutions Center
UNCED	United Nations Conference on Environment and Development
UNDP	United Nations Development Programme
UNFAO	United Nations Food and Agriculture Organization
USAID	United States Agency for International Development
UTM	Universal Transverse Mercator
VDC	village development committee

People and Forests

1

Explaining Deforestation: The Role of Local Institutions

Clark C. Gibson, Margaret A. McKean, and Elinor Ostrom

Introduction

Governments, citizens, and scientists are increasingly concerned about the role of forests in global environmental change. Evidence is mounting from multiple studies that humans at an aggregate level are exploiting forests at unsustainable rates in tropical regions.[1] While some deforestation can be attributed to rational and sustainable transfers of land to agricultural and other valuable uses, unplanned deforestation can generate significant negative externalities: loss of biodiversity, elevated risk of erosion, floods and lowered water tables, and increased release of carbon into the atmosphere associated with global climate change. Deforestation can also decrease the welfare of forest users by eliminating habitat for game species, altering local climates and watersheds, and destroying critical stocks of fuel, fodder, food, and building materials.

While aggregate levels of deforestation are relatively well known, less agreement exists among forest managers, policymakers, and scholars about the underlying and proximate causes of these increases.[2] The most frequently mentioned causes of deforestation include

- Population growth (Rudel, 1994),
- Population density (Burgess, 1992),
- Affluence (Ehrlich and Ehrlich, 1991; Rudel, 1994),
- Technology (Ehrlich and Ehrlich, 1991),
- National debt (Kahn and McDonald, 1994),
- Commercial logging (Capistrano, 1994),

- Government policy (Repetto and Gillis, 1988; World Bank, 1992),
- Forest accessibility (Kummer, 1992), and
- Political stability (Shafik, 1994).

Such disagreement about the most important factors means that there are multiple processes at work or that significant knowledge gaps exist about these processes or both. Even when agreement has been reached on the importance of a certain factor, researchers have disagreed about its effect. For example, while some researchers argue that population growth is a major cause of deforestation, Caldwell (1984) suggests there is no linear relationship between population pressures and land degradation. Bilsborrow and DeLargy (1991), as well as Wolman (1993), assert that solid empirical evidence about the impact of population pressure is almost nonexistent. In fact, Blaikie and Brookfield (1987) report that land degradation occurs in areas with both increasing and decreasing population pressure, and Allen and Barnes (1985) find no relationship between the population and deforestation. An important study by Tiffen, Mortimore, and Gichuki (1994) demonstrates the impact of a fivefold increase in population in the Machakos District of Kenya between 1930 and 1990. They provide substantial evidence that increased labor availability in the locality—when combined with market opportunities, technological knowledge, and appropriate institutions—has led to sustainable resource practices, including the planting and husbandry of more, rather than fewer, trees.[3] And Varughese (chapter 8 this volume) finds no direct link between population and deforestation in a comparison of 18 communities in the Middle Hills of Nepal.

Similarly complex and multidirectional results are reported for other variables asserted to be causes of deforestation, including

- Individual wealth (Shafik, 1994),
- National debt (Capistrano, 1994),
- Forest accessibility (Agrawal, 1995; Schweik, chapter 5 this volume), and
- Commercial logging (Burgess, 1992; Capistrano, 1994).

Contributing to such contradictory findings is the dearth of accurate forestry data at the national, regional, and local levels; the lack of time-series data; the lack of good institutional data; and the disparate definitions and measurements employed in studies of deforestation (Kaimowitz and Angelsen, 1998).

Additionally, many analyses of forest exploitation lack linkages to the local level, despite a growing awareness among scholars and practitioners that the actions of local people greatly determine the success or failure of schemes regarding natural-resource management.[4] Because much of the debate about the causes of deforestation and other environmental harms has been largely confined to macroanalyses, it has failed "to benefit from the wealth of data generated at the micro level—data that provide rich information on the social and economic factors that mediate the relation between population and the environment" (Arizpe, Stone, and Major, 1994, 3; but see Wollenberg and Ingles, 1998; Poffenberger and McGean, 1998). And even though there are numerous local-level studies of forests and their users, the number of studies with "careful, quantitative micro-level empirical research . . . is not impressive" (Kaimowitz and Angelsen, 1998, 99).

And yet the role of people at the local level is crucial. National governments rarely possess enough personnel or money to enforce their laws adequately, prompting many officials to consider decentralizing authority over forest resources. It is becoming increasingly clear that local communities both filter and ignore the central government's rules. They also add their own rules, generating local institutions—rules-in-use—and patterns of activity that can diverge widely from legislators' and bureaucrats' expectations. Because local communities live with forests, are primary users of forest products, and create rules that significantly affect forest condition, their inclusion in forestry-management schemes is now considered essential by many researchers and policymakers (Arnold, 1992).[5]

The authors in this volume seek to understand the complex interactions between local communities and their forests. To do so, they depart significantly from conventional national-level analyses and offer groundbreaking efforts to identify the relationship between forest conditions, individuals, and institutions at a local level. The presumption that guides the authors is that institutions at the local level—together with the incentives and behaviors they generate—lay at the heart of explanations of forest use and condition (Thomson, 1992).

Local institutions can modify the effect of factors thought to be the driving forces of deforestation (see, for example, Agrawal and Yadama, 1997). Rare is the market, technological, demographic, or political factor that affects individuals without first being filtered by local institutions. Given certain institutional arrangements, individuals may forgo the use of a resource if it is not culturally acceptable (see Schweik, chapter 5 this volume). Individuals may ignore central government rules that contradict their daily patterns of resource use (see Banana and Gombya-Ssembajjwe, chapter 4 this volume) or ask the central government for help in protecting their resources (see Agrawal, chapter 3 and Varughese, chapter 8 this volume). Individuals may construct rules to prevent the immediate commodification of their forest resource (see Agrawal, chapter 3, Becker and Leon, chapter 7, and Varughese, chapter 8 this volume) or they may allow the resource to be put on the market quickly (see Gibson and Becker, chapter 6 this volume). Since local institutions guide the daily consumption of natural resources, it is appropriate to keep them at the center of analyses concerning forest use.

Any analysis of how local institutions affect forest conditions necessarily crosses the neat boundaries of academic disciplines. Evaluating the condition of a forest requires employing the concepts and measurement techniques of biologists and ecologists. Understanding local behavior needs insights from anthropology and sociology. Examining the creation and enforcement of rules needs the input of political scientists. And estimating the impact of a forest on household budgets must borrow from the economists' toolbox. The authors of the empirical studies found in this volume invest substantial effort to weave together the natural and social sciences to create more comprehensive explanations of the people-forest nexus. Further, all of the cases explicitly use the methods of the International Forestry Resources and Institutions (IFRI) research program, which not only employs a multidisciplinary approach but allows for comparison across time and space as well (see the appendix to this volume).

Because the authors in this volume move away from simple, national-level studies of forests and toward more comprehensive accounts of forests and communities at the local level, their studies offer policymakers a more sophisticated view of forest management from which to derive policy options. The cases in this volume demonstrate that forests should not be considered as the source of only one commodity, wood; nor should users of the forest be clumped together as one group. Rather, these studies underscore how forests are associated with *multiple products* (for example, wood for construction and fuel, wildlife, water, leaves, fruits, fodder, seeds, straw, shade, fertile soil, stones, and so on) and *multiple user groups* (defined by property rights, product, location, citizenship, religion, caste, ethnicity, technology, income, and access). The variation of local institutions discovered by the authors also discourages the view that template forest policies are likely to work when imposed on a country as a whole. The diversity of conditions, rules, and outcomes presented in this volume's chapters, therefore, equips policymakers with an appreciation for the complexity of forestry resources as well as examples of management successes and failures that should assist in the construction of the most appropriate roles to be played by local, regional, and national authorities.

Forests, Goods, Rights, and Owners

Clarifying the differences and similarities between types of goods, property rights, and owners is an essential first step toward an understanding of the interaction between people and forests. McKean explores these concepts in chapter 2, noting that the differences between public and private types of goods, rights, and owners are more than semantic. The differences can have critical effects on the distribution of a forest's benefits and, ultimately, on the overall condition of forests. To misjudge the types of goods involved with a resource system can lead to the design of inappropriate property-rights arrangements, and these can in turn create the incentive for grievous depletion rather than sustainable use.

As economists have long defined these things, property rights to resources are not the resources themselves but are human institutions, sets of mutually recognized claims and decision-making powers over those resources. Private property rights are those that are clearly specified (not vague), secure (not subject to whimsical confiscation), and exclusive to the owner of the rights. Rights that are vague, tenuous, or nonexclusive are not fully private. Private property arrangements win praise and admiration, appropriately, because they can encourage protection and investment in the goods to which they attach. Of course, they cannot do this—perhaps nothing can—in an atmosphere of chaos, insecurity, and short time horizons, and we would be wrong to blame the property-rights institutions when the real problem is overwhelming uncertainty.

McKean argues that much of the theoretical foundation underpinning the debates over property rights assumes that there are only two kinds of goods: public goods and private goods. For several decades now, political economists have agreed that the two crucial dimensions we should use to classify goods are (1) the ease with which potential users can be excluded from access to the good (the *excludability* of the good) and (2) whether using a portion of the good shrinks the supply that remains (the *subtractability* or *rivalness* of a good). Pure public goods are nonexcludable and nonsubtractable, and private goods are both excludable and subtractable. The dichotomy of pure public goods and private goods has become the focus of discussion about types of goods ever since, and consequently many have overlooked the other two types of goods that are created by this two-by-two typology: club goods are excludable but nonsubtractable, and common-pool goods are difficult to exclude but subtractable. Little harm has been done by ignoring club goods because they are easy to produce (because they are excludable) and undepletable (because they are nonsubtractable). However, ignoring common-pool goods, which are difficult to produce and easy to deplete, is tragic. It turns out that most environmental and natural resources that we care about are common-pool goods. They are as subtractable as private goods, but because it is difficult to control or restrict access to them (the excludability dimension), it is very difficult to restrict the rate at which they are consumed. Thus, we arrive at a recognition of environmental crisis rather underequipped and ill accustomed to thinking about the crucial features of environmental resources. Because we have become accustomed to thinking in terms of only public goods and private goods, when we recognize that environmental resources are subtractable we begin to think of them as private goods.

If forests were like farms, producing wood as farms grow tomatoes or flax, then viewing them as private goods and creating individual private property rights in forests might be sensible. But even monoculture tree farms are frequently complex ecosystems of varied and interdependent species producing multiple products. Nonmonoculture forests are even more complex, generating goods that range from fallen leaves to berries to kindling to timber, and their resilience as productive systems requires that complexity. They also provide environmental services beyond the forest, in terms of erosion control, flood control, conservation of water, cleaning of air and water, and stabilization of local climate. The size of many forests, and the inevitable complications involved in monitoring the use of the forest and balancing one use against another, make exclusion or restrictions on access intrinsically problematic. Thus, McKean asserts that it is appropriate to think of forests as a complex of many commodities with attributes of both common-pool and public goods.

The definition of private property rights has to do with the clarity, security, and exclusivity of the right and does not actually include any stipulation that they be vested only in single individuals. Although larger entities and groups of individuals may theoretically hold private property rights—and do in actual fact as well (for example, business partnerships and joint-stock corporations)—much discussion forgets this. As a result, campaigns to create private property rights tend to consist of transferring ownership from larger entities and groups to individuals. In some instances, these interventions may destroy the property-rights arrangements that they should want most to create. Most privatization campaigns would ignore or even oppose the assertion that there might be conditions when it is more desirable for clear, specific, secure, and exclusive rights to be vested in a group rather than in single individuals, but McKean outlines conditions in which group rights may make more sense.

It is widely agreed that private property rights are the appropriate institution to create for commodities that are subtractable and from which it is easy to exclude others from benefits. Thus, if one thinks of natural resource systems as potentially private goods, one will advocate creating private property rights for those resources. And if one's notion of private property rights requires vesting all such rights in individuals, then one will fail to consider the possibility of vesting rights in groups or communities when that might be appropriate. McKean argues that natural-resource systems that are really combinations of public and common-pool goods can have as many as four attributes that make vesting property rights in groups more efficient than vesting those rights either in a single individual or trying to parcel the resource into individually titled patches.

First, some resources are simply indivisible, and some resource systems like forests contain or produce useful items that are themselves fugitive or mobile resources. Second, on some large resource systems, particularly in arid regions, there is great uncertainty in the location from year to year of the most productive zones. Third, on resource systems with congested and competing uses and high population pressure, coordination among users is essential to cope with externalities. Fourth, group ownership and thus group enforcement of rules can be an efficient way to cope with the costs of monitoring otherwise porous boundaries and enforcing restraints on use within those boundaries. In many resource systems including forests, more than one condition, or even all four conditions, may pertain. Thus, forests make good candidates for common-property regimes—or for vesting clear, specific, secure, exclusive rights to managing a resource in nearby communities.

The contributions in this volume address a variety of property-rights arrangements and take into consideration how the institutions that surround these arrangements provide incentives for local residents to use their forests. These property-rights arrangements often have critical influences over the condition of forests.

IFRI Research Program

The empirical chapters following McKean's theoretical exploration accept the challenge that our understanding of forests relies on our understanding of how people at the local level interact with forest resources. In their quest for untangling these complex relationships, the authors of these chapters draw on the design, principles, and hypotheses of the International Forestry Resources and Institutions (IFRI) research program. The IFRI research program is a multilevel, multicountry, overtime study of forests and the institutions that govern, manage, and use them.

To help explain deforestation and loss of biodiversity, the IFRI research program draws on the Institutional Analysis and Development (IAD) framework developed and used by colleagues associated with the Workshop in Political Theory and Policy Analysis at Indiana University over several decades (Kiser and Ostrom, 1982; Ostrom, 1986; Oakerson, 1992; Ostrom, Gardner, and Walker, 1994). The IAD framework has been used to study how institutions affect human incentives and behavior as these impact on urban services in metropolitan areas, the provision and production of infrastructure (such as roads and irrigation systems), and the governance and management of natural resource systems. At the core of the IAD framework are individuals who hold different positions (members of a local forest-user group; forest officials; landowners; elected local, regional, or national officials) who must decide on actions (what to plant, protect, harvest, monitor, or sanction) that cumulatively affect outcomes in the world (forest conditions, the distribution of a forest's benefits and costs). To simplify representation, the complex set of incentives and resulting behavior is initially represented in figure 1.1 as a single box. This "box," like all of the other boxes in figure 1.1, can be opened and contains a nested set of other conceptual boxes within it.

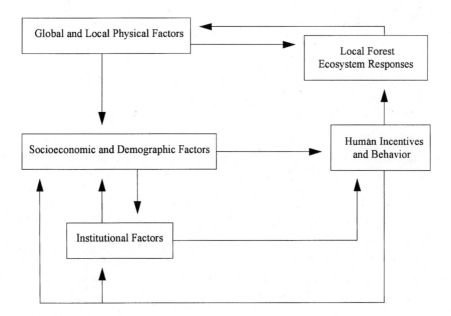

Figure 1.1
The IAD framework relating multiple factors affecting local ecosystems

In a dynamic setting, human behavior impacts local ecologies that are also affected by (and affect) global and local physical factors (including changing technology). Human incentives and behavior are affected by socioeconomic and demographic factors as well as institutional factors. Each of the factors on the left-hand side of figure 1.1 unpacks into a large set of variables. For example, unpacking the institutional factors that may affect human incentives and behavior across a large number of diverse settings includes variables at multiple levels. At a micro level, these would include, but not be limited to, such variables as

• Specific rules-in-use for each parcel of land (or forest product) in a local ecology that differ in regard to who can harvest, when and how, and how much harvesting of different products is authorized or forbidden;

• The types of afforestation or other enhancement or protection activities that are encouraged and by what means;

• The types of subsidies that are provided related to the inputs or outputs of a local economy;

• The methods of monitoring and sanctioning forest use and investment practices;

• The level of common understanding of what rules are used, monitored, and enforced;

• The degree of organization among forest users and the meaning of such organization in terms of individual incentives;

• The representatives of local, regional, or national government who are involved in local activities.

At a macro level, these would include, but not be limited to, such variables as

• National legislation authorizing diverse types of forests and parks in a country and the restrictions or subsidies involved in the use and administration of each type of forest;

• The types of private or communal land and tree tenure authorized;

• The personnel rules of national, regional, and local agencies affecting recruitment, retention, promotion, and discipline of public officials;

• Taxation laws on land, extraction rates, and corporate profits;

• The availability of courts to resolve disputes over land or tree tenure, contracts related to concessions, and disciplinary actions within public agencies.

Systematic information about institutional variables at a micro level are not available in any existing data set, nor are most relevant macro-institutional variables.

One advantage of a simple framework is that a large number of nested variables can be included. And, given the complexity of the forest–local community nexus, such complexity was a given. Workshop colleagues sought input from a wide range of international scholars, including biologists, ecologists, resource economists, foresters, anthropologists, sociologists, demographers, lawyers, geographers, and political scientists. Their input was even more deeply embedded after early field testing occurred in Bolivia, Nepal, and Uganda. Thus, researchers from a variety of disciplines contributed invaluable advice about the factors that may help explain how humans impact forest condition and biodiversity. Given these many and interrelated factors, Workshop colleagues also employed a relational database to record the information gleaned by the IFRI protocols and to allow the testing of a nearly unlimited number of specific hypotheses.

IFRI researchers have concentrated first on the design of 10 research protocols and careful field methods for collecting valid and reliable information about microlevel institutional, socioeconomic and demographic, and local physical factors that affect human incentives and behavior, and the impact of this behavior on local forest ecologies.[6] It is the first research program to our knowledge that combines systematic forest mensuration techniques for a sample of 1-, 3-, and 10-meter radius forest plots for each forest in sites where data is also systematically collected about local institutions and socioeconomic and demographic variables.

In the early stages of this research program, IFRI colleagues are analyzing a small number of cases from the initial countries where research has been conducted—Bolivia, Ecuador, India, Nepal, Uganda, and the United States. The analyses contained in this volume, for example, range from a focus on a single case study to as many as 18 cases. All of the individual studies, however, have utilized the same research protocols. Thus, as the number of studies within each country grows, it will be possible to analyze results from an ever larger number of sites. Further, IFRI researchers intend to revisit sites on a regular basis to investigate more precisely the dynamics of how local institutional changes impact on the actions of forest users and officials as well as the results of these actions on forests. Thus, the IFRI research program provides a unique opportunity to undertake systematic, micro-level, comparative studies of institutions and their impact on rates of deforestation over time.

This volume represents our initial effort to report on studies conducted in Bolivia, Ecuador, India, Nepal, and Uganda based on a common framework and using the same research protocols. Since the IFRI research program has just entered its operational phase, we hope this is the first of a growing series of publications helping policymakers, forest users, and scholars understand the microprocesses at work under the macrovariables that have been the focus of recent attention.

Empirical Chapters

The empirical studies in this volume seek to fill at least two critical gaps in current forestry research. The first is the lack of comparable micro-level studies. The second is the shortage of studies that address the pivotal influence of local-level institutions on forest use and condition.

Micro and Comparative Analyses Are Important Because of Variation at Local Levels

Country-level data on rates of deforestation do little to help policymakers and scholars unravel the web of the causes of forest use. For example, while Uganda and Nepal have the same rate of deforestation at the national level, around 1 percent, these deforestation rates vary significantly within each country over space and time (FAO, 1993). And yet for forestry policy to be effective, an understanding of the causes of such dynamic and spatial variation within a country is critical. The empirical studies in this volume clearly demonstrate the need for scholars and policymakers to appreciate such local-level variation.

In chapter 3, Agrawal investigates how local-level variation within the Indian forest council system of community forestry leads to substantially

different outcomes for the management of forest resources. Agrawal begins his analysis by reviewing the legislation that undergirds the council
system. In response to widespread protest to the confiscation of lands by
the colonial government, the British passed the 1931 Van Panchayat Act,
which allowed village communities to create councils to control forested
areas previously administered by state revenue officials. While the Act
includes the broad outlines of the council's powers, local factors still generate the pattern of a council's day-to-day operations.

Agrawal demonstrates that these local factors help to explain why not
all of the councils have managed their forest resources successfully. Comparing nine councils from the Pithoragarh and Almora Districts, operating within the same ecological and administrative areas, Agrawal
finds that the councils range widely in terms of their size, organization,
age, and resource endowments. Evaluating how these characteristics affect forest condition, Agrawal argues against those who would assert that
either per capita income or the age of councils are the major factors that
account for the success of local councils in managing their forest resources. Rather, Agrawal indicates that the size of the council is an ignored but important factor that affects its performance. Very small
councils are disadvantaged, Agrawal argues, in their efforts to generate
sufficient human and other resources to monitor and enforce local rules.
Moderate-size councils are able to generate greater amounts of monetary
and voluntary contributions in their efforts to monitor the use of their
forests, which are under constant threat of exploitation by locals and
outsiders. These findings challenge those scholars and practitioners captured by an invariant "smaller is better" view. Rather, Agrawal indicates
that somewhat larger organizations can have great advantages in managing forest resources at the local level. Additional studies of councils are
planned that will enable Agrawal to examine a broader array of these
local institutions so that the possibility of a curvilinear relationship between size of forest organization and capabilities to monitor and enforce
local rules can be explored.

Banana and Gombya-Ssembajjwe's analysis of forests in Uganda (chapter 4) further underscores the diversity of outcomes at the local level. In
their examination of five forests located in four different ecological zones,
Banana and Gombya-Ssembajjwe discover that the level of human

consumptive activity differs widely and has a dramatic impact on the physical condition of the forests. Three forests (Mbale, Lwamunda, and Bukaleba) show signs of heavy use in the forms of illegal commercial logging activities and livestock grazing; over 70 percent of the 90 sample plots had evidence of illegal utilization. Two other forests (Namungo and Echuya), however, showed significantly less disturbance, despite the fact that they, too, contain valuable commodities such as commercial tree species and grazing areas.

Discounting environmental and biological factors as explanations for this variation, Banana and Gombya-Ssembajjwe then consider social explanations. They indicate that most forested lands in Uganda are state property, thus offering little incentive for locals to constrain their consumption of forest products. Colonial and postcolonial regimes vested forested lands within the central government, disregarding indigenous property rights or management schemes. Without a stake in the tenure of the resource, the authors argue, local villagers have the incentive to consume forest commodities opportunistically. Thus, the degradation of Uganda's forested lands should be expected.

But Banana and Gombya-Ssembajjwe assert that this general lack of tenure at the local level does not explain the variation of forest condition found in their five cases. The authors turn to the level of enforcement for each forest to account for these differences. Mbale, Lwamunda, and Bukaleba Forests are all state-owned forest reserves. Each forest is monitored only by Uganda's Forest Department, which possesses relatively few staff to fulfill their protective function. Further, Department staff have few incentives to patrol frequently, since the benefits resulting from their employment are not closely tied to their enforcement of the law. During the past several decades, the Forest Department has not been able to enforce its rules in a uniform manner. Thus, little common understanding exists of what rules might actually be in practice. The Echuya and Namungo Forests, on the other hand, both have had a much greater stability in the rules that are enforced and a much greater level of monitoring and enforcement. While Echuya is a government reserve, the Forest Department has augmented its monitoring capabilities by using the help of an Abayanda (pygmy) community that resides in the forest. The Abayanda benefit from access to forest products in return for their monitoring duties. Namungo's Forest is a privately owned woodland for which a family hires its own guards. Those villagers who live near to Namungo's Forest also help monitor its use since the family allows villagers their traditional rights to extract certain levels of firewood, poles, medicines, fruit, fodder, and other forest products. Thus, Banana and Gombya-Ssembajjwe demonstrate that property rights and their enforcement help to explain the variation of forest conditions found in their site.

Schweik's analysis in chapter 5 delves even more deeply into issues regarding the geographic variation of forest condition. Schweik seeks to account for the spatial variation of the Sal tree, *Shorea robusta,* that villagers living in the Chitwan District of southern Nepal find particularly valuable for fuelwood, tool-making, and construction. Schweik sets out to test three rival hypotheses: (1) Sal exhibits a pattern of natural regeneration where human disturbance is not detected, (2) Sal exhibits a pattern of optimal foraging, and (3) Sal exhibits a pattern of optimal foraging altered by institutional and social norms that exist in the area.

Using a sophisticated combination of tools including global positioning system (GPS) equipment, geographic information system (GIS) software, the IFRI research protocols, and a maximum-likelihood regression model, Schweik builds three imbedded regression models to capture the influence of important factors that affect the growth pattern of Sal and to test each hypothesis.

To establish the human and nonhuman impedances to the growth of Sal, Schweik first gathers data from a relatively undisturbed forest to establish the unimpeded or "natural" distribution of Sal. In such a setting, the tree lives in clusters, generating a negative binomial distribution (as opposed to a random or uniform distribution of trees), a finding critical to the appropriate specification of the statistical models.

Schweik's results reveal that slope steepness is a critical factor in determining where Sal trees are found. The author also finds that two spatial variables—the elevation and the east-west location of plots—to be significant, and he links them with human behavior at the local level. Given that Sal grows at elevations up to 1,200 meters, its distribution should not be affected in the area under study (extant hills do not exceed 800 m). Schweik's results, however, show that in the study site, the number of trees increases at higher elevations. Such an outcome resonates with optimal foraging theory, which argues that individuals seek the easiest source for their resources: climbing hills to gather trees makes them more difficult to acquire, and thus fewer would be taken at higher elevations. This evidence supports Hypothesis 2. However, the decrease in trees from west to east, however, is not captured by either the nonhuman factors or simple optimal-foraging theory, since the pattern of exploitation should result only in a ringed pattern surrounding villages, not in a systematic decrease in trees from west to east. This result supports Hypothesis 3. Schweik finds the operation of Nepal's caste system coupled with more effective monitoring of harvesting rules by forest guards in the West to be the most convincing explanation for the west to east decrease of Sal. Institutional and social factors appear to be shifting where foraging has occurred. In other words, patterns of optimal foraging are altered due to the structure of the institutional landscape. Thus, the forests of the east are being used at a greater rate than those in the west. Schweik's path-breaking analysis demonstrates how human use patterns vary significantly at the micro level, leading to differences in forest condition within forested areas as small as 10 square kilometers.

Gibson and Becker's examination of the relationship between the members of the Loma Alta commune and their fog forest in Ecuador highlights how the nexus of users, property rights, and forest products may account for the variation found in a forest's condition (see chapter 6). Their study of the *comuna* is timely: Loma Alta is one of many comunas located along the watersheds of the Chongon Colonche mountain range of western Ecuador, whose last stands of tropical forest are home to numerous endemic species—so many, in fact, that some conservationists consider the area's protection a global priority.

Unlike other national governments—and central to this study—is the fact that Ecuador recognizes the rights of some local communities to govern their local affairs. In 1936, the central government passed the Law of the Comunas, empowering 32 communities living in the coastal areas to hold land jointly and act as their own local governments. Although the land is held in common, the comuna still allocates its members distinct plots to use as they see fit. The members' rights to the land are constrained by only two rules: they must use the land, and they may not sell it. Otherwise, the plots are treated as private property, with members making capital improvements to the land, passing it on to their offspring, and renting it to other comuna members.

Gibson and Becker argue that this system of property rights directly affects the condition of the comuna's upland fog forest. In the part of the forest that has not been allocated to individuals, members and outsiders have seriously degraded the forest. Approximately 70 percent of the forest cover has been removed, and large cleared areas exist—testimony to the commercial selling of timber and the conversion of forest into pastureland. Where individuals have been allocated plots in the forest, however, it has endured far less exploitation.

Variation also exists within those plots that have been allocated to individuals. At elevations above 300 m, some land within the forest has been cleared to establish plantations of the cash crop paja toquilla (*Carludovica palmate*). Farmers plant paja at this elevation since the tree needs the moisture that the forest at higher elevations provides.

Gibson and Becker find that the particular system of property rights within the Loma Alta comuna, the value of the forest as land for paja toquilla, and timber sales has led to a specific pattern of deforestation in the communal forest. Although many parts of the forest still display the characteristics of a relatively healthy secondary forest (having been commercially logged over the last century), the authors argue that the forest remains threatened by the possible expansion of farming activities and the lack of comuna rules regarding land use.

Becker and León (chapter 7) investigate the variation that occurs in forest conditions even where used by the same ethnic group along the same river. The authors focus on the relationship between three Yuracaré settlements and their adjacent riparian forest along the Rio Chapare in Bolivia. In their attempt to explore if and how these Yuracaré communities might manage their forest, Becker and León draw on biological measures of the forests and compare them with a reference forest of the same type that is known to have been relatively unused. In addition, the authors selected the three sites because they vary in their distance from the closest market and in their population.

Becker and León find a complex pattern of behavior and outcomes in their study. The forests do, in fact, display predictable variations along the dimensions of moisture gradient, distance to markets, and population pressure. But the authors find results that go beyond these simple causes, the most important of which is that the Yuracaré are clearly managing their forested areas to increase the populations of game animals. By planting and tending to fruit trees, the Yuracaré intentionally alter the forest to suit their preferences for certain food types. Becker and León argue that these local institutions are under threat, however, as markets increasingly penetrate the area, causing changes in Yuracaré preferences in food and labor.

In chapter 8, Varughese encounters substantial variation in both the condition of the forests and community management in his study of 18 cases in the Middle Hills of Nepal. Although all the communities he studied depended on forest products to substantial degrees, Varughese found that forests ranged from being in good condition (as evaluated by a professional forester along both subsistence and commercial scales) to poor condition. In some sites, this condition was improving; in others, it was growing worse.

Varughese, interested in explaining this variation, first tests the simple hypothesis that population (measured in different ways) drives variation in forest condition. This view was widely held in the 1970s and 1980s, especially when several studies showed alarming patterns of deforestation in Nepal. This simple neo-Malthusian approach would argue that in those locations where population is large or growing, the forest would be put under additional pressure, leading to its worsening condition. In those sites where population is low or steady, on the other hand, forest condition should be good or stable.

Varughese, however, finds no support for this argument. In his sample, areas of high population contain forests of both good and poor condition, as do areas of low population. Since population does not appear as a major driver of forest depletion (and variation) in his sample, Varughese investigates the role of local institutions. He finds that those communities that have a higher level of organization regarding the forest—as measured by the presence of institutional arrangements such as monitoring assignments and restrictions on entry and harvesting—tend to have forests in better condition.

Micro and Comparative Analyses Are Important Because of Institutional Variation

The micro and comparative studies of deforestation found in this volume do more than merely offer data regarding local-level variation; they offer much-needed analyses of the workings of local institutions as well. For most of the authors, institutional environments emerge as a critical factor in accounting for a forest's condition. The authors show how local forest users are able in some cases to devise rules regulating access and use that reduce the pressure to overharvest. Thus, there is substantial innovation and creativity exhibited in many settings. On the other hand, the task of devising adequate rules to govern and manage forest resources is particularly challenging and one that is not always achieved by local forest users or by central government officials. Degraded forest landscapes have resulted in spite of pronouncements made by officials and the best intentions of local forest users. It is, thus, particularly important to learn from both successful and unsuccessful local cases what factors tend to account for successful development of local institutions that enhance forest conditions.

Using data from the forest-council system of India, Agrawal challenges the current conventional idea that smaller groups manage their resources better than larger groups. The lesson one learns from his chapter is that forest users who develop successful local institutions must contribute time and effort to monitor and enforce the rules they have crafted for their own setting. Self-governing institutions are costly, especially when they regulate a territory that is relatively large and not immediately visible to local villagers as they go about their daily tasks. A moderate-size village appears able to generate the time and resources needed to control access to a forest that a very small village cannot and to avoid the costs of organization that may plague larger villages.

Schweik demonstrates that physical variables alone do not account for differences in the availability of a valuable tree species. There is, however, a subtle institutional factor that does help to account for the availability of these trees. The differential access of high-caste forest users, as contrasted to low-caste forest users, provides an explanation for the variation of these trees across forest space. Without the statistical analyses conducted by Schweik and data regarding institutional variables, it would have been difficult to sort out the relative importance of the many physical and cultural factors at play in the villages' forests.

Banana and Gombya-Ssembajjwe also demonstrate the subtle differences among institutions at the local level of forest use. In their study of one private forest and four government forests, they identify two forests where rule conformance is generally much higher. Even though these two forests have quite different formal structures, the time horizons and immediate incentives of participants are such that monitoring rule conformance occurs at a higher rate. The physical structure of both forests reduces the time and effort needed to achieve higher levels of rule conformance. The lessons we learn from this chapter reinforce Agrawal's analysis and the importance of understanding how physical variables and locally understood and enforced rules and norms jointly affect incentives and behavior.

Even with appropriate-size groups and security of tenure, however, successful resource management may not occur. A great many development scholars and practitioners aver that microinstitutional arrangements are likely to be created in an environment where local autonomy is tolerated, where a history of institutional creation has occurred, and where local communities have secure property rights. In their study of a rural community in western Ecuador, Gibson and Becker find most of these institutional prerequisites for successful natural resources management exist, and yet the authors discover parts of the community's tropical forest characterized by open access. The authors argue that even within a local community, differences between user groups and the variation over their preferred forest product critically affect how and if rules are created to manage forest resources.

Becker and León's study of the Yuracaré challenges those in the central government of Bolivia who had thought forested areas of the Amazon were unmanaged. The Yuracaré have a long history of managing their forests for particular ends. The authors find evidence of such forest institutions in the language of the Yuracaré as well as in the biological condition of the forest where indigenous timber species are more conserved than commercial timber species, and fruit trees preferred by the game the Yuracaré hunt are planted and nurtured.

Locally constructed institutions are at the center of Varughese's explanation of forest condition in 18 sites in Nepal as well. In those sites where communities have crafted institutions to deal with the management of forest resources, the forest tends to be in better condition than in those sites where communities have not made attempts to, or confronted obstacles to, efforts at organizing themselves. Varughese finds that such obstacles can result from both internal and external sources. This study offers powerful evidence that research that focuses solely on population as a driver of deforestation may be far off the mark, especially if the attempt is to explain the variations that may be found at the local level.

Conclusion

By featuring variation at the local level, this volume offers a general lesson to policymakers interested in forest management: national- or even regional-level policy may not fit local circumstances. The studies show that within even relatively small, ecologically similar areas under the same set of national laws, numerous nonbiological factors help to explain variation in forest condition. Different user groups, systems of property rights, types of commodities taken from a forest, and extant levels of rule enforcement interact with national legislation in different ways to produce particular patterns of forest use and conditions. Thus, while each local community operates under the same national legislation, their behavior and impact on forests differs substantially. For example, Agrawal and Varughese both report that some local communities respond by hiring guards to protect their forests while others do not. Banana and Gombya-Ssembajjwe demonstrate that locals enforce national forestry legislation in some areas of Uganda while in other areas it is ignored by community members. Schweik claims that most individuals in his study area routinely flout the national law proscribing wood harvesting. Such cases reveal that forest management is intensely local and that national legislation can be modified, ignored, or enforced by local communities to fit their circumstances.

In addition to the lessons generated by the cases' variance, they also offer common insights regarding how management schemes may be successful. One crucial factor that emerges is the importance of commonly understood rules and their enforcement. Successful enforcement at the local level partially depends on individuals who generally agree on what rules they should follow (and, hopefully, why they have been adopted). Without this agreement, there is less incentive to comply with rules: if either local forest users or government guards monitor forest use, a lack of agreement about rules would achieve a lower level of rule compliance. Efforts to guard effectively in this case result either in the type of corruption that often occurs between government guards and local forest users (especially bribery) or very high levels of conflict. Once some common agreement is achieved, then investment in monitoring has a high return by ensuring that the temptations that face all users do not grow into consistent rule-breaking behavior. In the case from Uganda, for example, the well-understood and long-standing extension of traditional rights by a private owner to nearby residents combined with active monitoring has generated a forest in relatively good condition, especially as compared to a neighboring government forest that does not enjoy much protection from its government guards. One of the central points of Agrawal's investigation is that moderate-size communities that agree on a general set of rules regarding forest use can better afford to share monitoring duties and thus enjoy better forest resources. Gibson and Becker find that lands that lack an agreed-upon set of rules for their use are overexploited by both locals and outsiders. These studies concur with the growing theoretical consensus that argues that without common understanding and resources sufficient to monitor and sanction rule breakers, rules restricting activities that generate high private benefits are moot, whether made and enforced by the national government or by the local community.

Notes

1. In contrast, the area and volume of forest resources are growing in most temperate regions.

2. For a brief overview of the competing explanations given for deforestation, see Turner (1995).

3. What is important about the Tiffen, Mortimore, and Gichuki study is that it demonstrates the variability of responses to population changes in different lo-

calities. It challenges the presumption that a population increase at a local level will harm the ecological system at the local level. It does not address the question of population increases at a global scale (see Holling, 1994, for an overview of ecological research showing diverse responses at multiple scales to population increases), nor does it address the issues regarding secondary forests that may result from human efforts to restore areas where primary forests previously stood.

4. See, for example, Arnold (1998), Ascher (1995), Berkes and Folke (1998), Blockhus et al. (1992), Bromley et al. (1992), UNFAO (1990), Gibson and Marks (1995), Hecht and Cockburn (1990), Marks (1984), McCay and Acheson (1987), Ostrom (1990), Poffenberger (1990).

5. This realization of the importance of the local has led both governments and scholars to examine comanagement options (see Lynch and Talbott, 1995). Several of the chapters in this volume include cases that may be referred to by some as *comanaged*. See in particular chapters 5 and 8. While most of these chapters do not directly address this option, future work will more directly discuss comanagement as a means of governing large forests (as contrasted to the smaller forests that are the focus of this book).

6. Now that the design of the micro-level studies has been completed, we are starting to design a macro-level study using the same framework but including variables characterizing national-level entities.

References

Agrawal, Arun. 1995. "Population Pressure = Forest Degradation: An Oversimplistic Equation?" *Unasylva* 46(2): 50–58.

Agrawal, Arun, and Gautam Yadama. 1997. "How Do Local Institutions Mediate the Impact of Market and Population Pressures on Resource Use?" *Development and Change* 28(3): 435–65.

Allen, Julia C., and Douglas F. Barnes. 1985. "The Causes of Deforestation in Developing Countries." *Annals of the Association of American Geographers* 75(2): 163–84.

Arizpe, Lourdes, M. Priscilla Stone, and David C. Major. 1994. *Population and Environment: Rethinking the Debate*. Boulder, CO: Westview Press.

Arnold, J. E. M. 1992. *Community Forestry: Ten Years in Review*. Rome: FAO.

———. 1998. *Managing Forests as Common Property*. Rome: FAO Forestry Paper 136.

Ascher, William. 1995. *Communities and Sustainable Forestry in Developing Countries*. San Francisco: ICS Press.

Berkes, Fikret, and Carl Folke, eds. 1998. *Linking Social and Ecological Systems: Management Practices and Social Mechanisms for Building Resilience*. Cambridge: Cambridge University Press.

Bilsborrow, R., and P. DeLargy. 1991. "Landuse, Migration and Natural Resource Degradation: The Experience of Guatemala and Sudan." In *Resources, Environment, and Population: Present Knowledge, Future Options,* ed. K. Davis and M. Bernstam, 125–47. New York: Oxford University Press.

Blaikie, P., and H. Brookfield. 1987. *Land Degradation and Society.* London: Methuen.

Blockhus, Jill M., Mark Dillenback, Jeffrey A. Sayer, and Per Wegee. 1992. *Conserving Biological Diversity in Managed Tropical Forests.* Gland, Switz.: International Union for the Conservation of Nature.

Bromley, Daniel, David Feeny, Margaret McKean, Pauline Peters, Jere Gilles, Ronald Oakerson, C. Ford Runge, and James Thomson, eds. 1992. *Making the Commons Work: Theory, Practice, and Policy.* San Francisco: ICS Press.

Burgess, J. C. 1992. *Economic Analysis of the Causes of Tropical Deforestation.* Discussion Paper No. 92-03. London: London Environmental Economics Centre.

Caldwell, John. 1984. "Desertification: Demographic Evidence, 1973–83." Occasional Paper No. 37, Australian National University.

Capistrano, Ana Doris. 1994. "Tropical Forest Depletion and the Changing Macroeconomy, 1967–1985." In *The Causes of Tropical Deforestation,* ed. Katrina Brown and David W. Pearce, 68–85. Vancouver: UBC Press.

Ehrlich, Paul R., and Anne H. Ehrlich. 1991. *Healing the Planet: Strategies for Resolving the Environmental Crisis.* Reading, MA: Addison Wesley.

Gibson, Clark, and Stuart Marks. 1995. "Transforming Rural Hunters into Conservationists: An Assessment of Community-Based Wildlife Management Programs in Africa." *World Development* 23(6): 941–57.

Hecht, Susanna, and Alexander Cockburn. 1990. *The Fate of the Forest: Developers, Destroyers, and Defenders of the Amazon.* New York: Harper.

Holling, C. S. 1994. "An Ecologist View of the Malthusian Conflict." In *Population, Economic Development, and the Environment,* ed. K. Lindahl-Kiessling and H. Landberg, 79–103. New York: Oxford University Press.

Kahn, J., and J. McDonald. 1994. "International Debt and Deforestation." In *The Causes of Tropical Deforestation: The Economic and Statistical Analysis of Factors Giving Rise to the Loss of the Tropical Forests,* ed. K. Brown and D. W. Pearce, 57–67. Vancouver, British Columbia: UBC Press.

Kaimowitz, David, and Arild Angelsen. 1998. *Economic Models of Tropical Deforestation: A Review.* Bogor, Indonesia: Center for International Forestry Research.

Kiser, Larry L., and Elinor Ostrom. 1982. "The Three Worlds of Action: A Metatheoretical Synthesis of Institutional Approaches." In *Strategies of Political Inquiry,* ed. Elinor Ostrom, 179–222. Beverly Hills, CA: Sage.

Kummer, David. 1992. *Deforestation in the Postwar Philippines.* University of Chicago Geography Research Paper No. 234. Chicago: University of Chicago Press.

Lynch, Owen J., and Kirk Talbott. 1995. *Balancing Acts: Community-Based Forest Management and National Law in Asia and the Pacific.* Washington, DC: World Resources Institute.

Marks, Stuart. 1984. *The Imperial Lion: Human Dimensions of Wildlife Management in Central Africa.* Boulder, CO: Westview Press.

McCay, Bonnie J., and James M. Acheson. 1987. *The Question of the Commons: The Culture and Ecology of Communal Resources.* Tucson: University of Arizona Press.

Oakerson, Ronald J. 1992. "Analyzing the Commons: A Framework." In *Making the Commons Work: Theory, Practice, and Policy,* ed. Daniel W. Bromley et al., 41–59. San Francisco: ICS Press.

Ostrom, Elinor. 1986. "An Agenda for the Study of Institutions." *Public Choice* 48: 3–25.

———. 1990. *Governing the Commons: The Evolution of Institutions for Collective Action.* New York: Cambridge University Press.

Ostrom, Elinor, Roy Gardner, and James Walker. 1994. *Rules, Games, and Common-Pool Resources.* Ann Arbor: University of Michigan Press.

Poffenberger, Mark, ed. 1990. *Keepers of the Forest: Land Management Alternatives in Southeast Asia.* West Hartford, CT: Kumarian Press.

Poffenberger, Mark, and Betsy McGean, eds. 1998. *Village Voices. Forest Choices. Joint Forest Management in India.* New Delhi: Oxford University Press.

Repetto, Robert, and Malcolm Gillis, eds. 1988. *Public Policies and the Misuse of Forest Resources.* Cambridge: Cambridge University Press.

Rudel, Thomas. 1994. "Population, Development and Tropical Deforestation: A Cross-National Study." In *The Causes of Tropical Deforestation,* ed. Katrina Brown and David W. Pearce, 96–105. Vancouver: UBC Press.

Shafik, Nemat. 1994. "Macroeconomic Causes of Deforestation: Barking Up the Wrong Tree?" In *The Causes of Tropical Deforestation,* ed. Katrina Brown and David W. Pearce, 86–95. Vancouver: UBC Press.

Thomson, James T. 1992. *A Framework for Analyzing Institutional Incentives in Community Forestry.* Rome: FAO, Community Forestry Note No. 10.

Tiffen, Mary, Michael Mortimore, and Francis Gichuki. 1994. *More People, Less Erosion. Environmental Recovery in Kenya.* New York: Wiley.

Turner, Paul. 1995. "Explaining Deforestation: A Preliminary Review of the Literature." Working Paper. Workshop in Political Theory and Policy Analysis, Indiana University, Bloomington.

United Nations Food and Agriculture Organization (UNFAO). 1990. *Interim Report on the Forest Resources Assessment 1990 Project. Item 7 of the Provisional Agenda, Committee on Forestry, Tenth Session, 24–28 September, 1990.* Rome: FAO.

———. 1993. *Forest Resources Assessment 1990: Tropical Countries.* FAO Forestry Paper 112. Rome: FAO.

Wollenberg, Eva, and Andrew Ingles, eds. 1998. *Incomes from the Forest. Methods for the Development and Conservation of Forest Products for Local Communities.* Bogor, Indonesia: Center for International Forestry Research.

Wolman, M. 1993. "Population, Land Use, and Environment: A Long History." In *Population and Land Use in Developing Countries,* ed. C. Jolly and B. Torrey. Washington, DC: National Academy Press.

World Bank. 1992. *World Development Report.* New York: Oxford University Press.

2

Common Property: What Is It, What Is It Good for, and What Makes It Work?

Margaret A. McKean

For more than a decade now, I have been involved in the study of *common-property regimes* for forests and other natural resources—or what might be described as institutional arrangements for the cooperative (shared, joint, collective) use, management, and sometimes ownership of natural resources. Given this definition, common-property regimes should range from communal systems of resource use among hunter gatherers, to mixed systems of, for example, communal woodland with individually owned arable fields, all the way to gigantic collective farms in socialist economies, and even, for that matter, to community and other broadly shared rights to regulate the environmental consequences of individual behavior in industrial economies. However, although policymakers have picked up on the importance of property rights in affecting environmental outcomes, they are currently designing radical changes in property-rights arrangements in transitional economies with virtually no knowledge of the specifics of what we are learning about common-property regimes for natural resources.

Privatization of property rights is a global fad right now: privatization of public enterprise in capitalist countries, decentralization of control over public enterprises (that nonetheless remain publicly owned) in socialist countries, and privatization of property rights in general in post-socialist countries. In the developing world (which is largely capitalist), there is also great enthusiasm for the privatization of traditional community lands and some government-owned lands. I am in basic agreement with the objectives of this conversion: to increase efficiency (when is wasting human effort or natural resources ever justifiable?), to enhance the incentives for investment, and, most crucially in the case of environmental resources, to create the incentive for resource protection and sustainable management. But at the same time, I fear that this privatization, particularly of forest resources, is being conducted without sufficient consideration of such issues as these:

• In whom (to individuals or to groups, to how many persons, to which persons, with what distributional consequences) should property rights be vested?

• Which rights should be transferred—full ownership with rights of transfer, or just use rights?

• What kinds of resources should be privatized? Are all objects equally divisible? Should ecosystem boundaries matter?

This chapter responds to these questions by exploring what common property is, then itemizing some of the potential advantages of using common-property regimes to govern and manage environmental resources in general and forests in particular, and concluding with a short summary of what we already know about the attributes of successful common-property regimes.

Definitions

Common-Pool Resources
Before one can talk about what value may be derived from common-property arrangements, we need to define terms, particularly because of the long history of confusing and conflicting usage in this field. Our first task is to distinguish between types of goods (see table 2.1). *Common-pool resources* are goods that can be kept from potential users only at great cost or with difficulty but that are subtractable in consumption and can thus disappear. If we devise no way to exclude noncontributing beneficiaries from common-pool resources, these resources are unlikely to elicit investments in maintenance or protection, as is also true of pure public goods with which they share this trait of nonexcludability. Common-pool goods are also subtractable in consumption (like private goods), which means they can be depleted. Without institutional mechanisms that address excludability and subtractability, then, common-pool

Table 2.1
Type of good, by physical characteristics

	Exclusion Easy	Exclusion Difficult or Costly
Subtractable (rivalrous in consumption)	Private goods (trees, sheep, fish, chocolate cake)	Common-pool goods (forest, pasture, fishery, any environmental sink over time)
Nonsubtractable (nonrivalrous in consumption)	Club or toll goods (Kiwanis club camaraderie, festive atmosphere at a party)	Pure public goods (defense, TV broadcasts, lighthouse beams, an environmental sink at a given instant, a given level of public health, a given level of inflation)

resources are essentially open-access resources available to anyone—very difficult to protect and very easy to deplete.

Many goods once described as pure public goods (nonsubtractable in consumption) in economics textbooks—air, water, roads, bridges—really are not pure public goods at all. They are, in fact, subject to crowding, wear, and depletion. Although a great deal of theoretical work and experimental economics have been done on pure public goods, the truly problematic category, into which natural resource systems and environmental resources fall, is common-pool goods. There is some risk that we might extract overly optimistic lessons from theoretical and experimental work that actually concerns nondepletable pure public goods. Fortunately, new game-theoretic and experimental work based on common-pool goods is also being done (see Ostrom, Gardner, and Walker, 1994).

Common-Property Regimes
The nature of a good is an inherent physical characteristic, not susceptible to manipulation by humans.[1] But property institutions are human inventions. There is tremendous confusion in the historical use of the term *common property,* as we will see below, but I use it to refer to property (not nonproperty) that is common, or shared. Thus a *common-property*

regime is a property-rights arrangement in which a group of resource users share rights and duties toward a resource.

Oddly, the term *common property* seems to have entered the social science lexicon to refer not to any form of property at all but to its absence—nonproperty or open-access resources to which no one has defined rights or duties (Gordon, 1954; Scott, 1955; Demsetz, 1967; Alchian and Demsetz, 1973). The inefficiencies and resource exhaustion to which open-access arrangements are prone are well known.[2] Open access is an acceptable method for resource management only when we need not manage resources at all: when demand is too low to make the effort worthwhile. In a common-property arrangement, on the other hand, particular individuals share rights to a resource. Thus there is property rather than nonproperty (rights rather than the absence of rights), and these are common not to an infinite all but to a finite and specified group of users. Thus common property is not access open to all but access limited to a specific group of users who hold their rights in common (Runge, 1981, 1984, 1992; Bromley and Cernea, 1989; Bromley et al., 1992). Indeed, when the group of individuals and the property rights they share are well defined, common property should be classified as a shared private property—a form of ownership that should be of great interest to anyone who believes that private property rights promote long time horizons and responsible stewardship of resources.

Goods, Rights, and Owners

I am convinced that part of our confusion in usage of terms is semantic: we use the same pair of adjectives, *public* and *private,* as labels for three different pairs of things. We use them to distinguish between two different kinds of goods (public goods and private goods), between two different kinds of rights (public rights and private rights), and between two different kinds of bodies that may own things (public entities or governments, and private entities or firms and individuals). Economists have for decades agreed that the privateness of a *good* is a physical given having to do with the excludability and subtractability of the good and that these two attributes of a good are crucial to understanding what humans can and cannot do with different kinds of goods. This definition of goods, creating the four-way typology shown in table 2.1, goes virtually unchallenged, although it is sometimes forgotten or misused.[3] The privateness of a *right* refers to the clarity, security, and especially the exclusivity of the right: a fully private right specifies clearly what the rights-holder is entitled to do, is secure so that the holder of the right is protected from confiscation by others, and is exclusively vested in the holder of the right and definitely not in nonholders of the right. It is important to note here that the privateness of a right has to do with the right and not the entity holding it; there is no requirement that this entity be a single individual. Finally, the privateness of a *body* has to do with its representational claims, in that a *public body* claims to represent the general population and not just one interest within that population, whereas a private body represents only itself.[4]

This confusion of the publicness and privateness of *goods* (a natural given), *rights* (an institutional invention), and *owners* of rights (entities that make different representational claims) has led to serious errors. First, we get *goods and owners* mixed up, falling very easily into the habit of thinking (1) that public entities own and produce public goods while private entities own and produce private goods and (2) that anything produced by government is a public good and anything produced by private parties is a private good. In fact, of course, there is no intellectual reason for this simple pairing off. Public entities are perfectly capable of producing private goods, and private entities occasionally produce public goods (though not often intentionally). Second, we get *goods and rights* mixed up and often attempt to create public rights in private goods and private rights in pure public goods or common-pool goods, with tragicomic effects (such as awarding an infinite number of rights to an exhaustible resource or awarding exclusive rights to resources that cannot be exclusively held). Third, we get *rights and owners* mixed up, thinking that private entities hold private (exclusive) rights and public bodies hold public rights. In fact, public rights (rights of access and use that do not include the right to exclude others from such use) are generally held by private entities because public bodies have awarded such rights to citizens. Similarly, public bodies hold both public rights (say, the use of an assembly hall or a courtroom that is also open to all citizens as observers) as well as private rights (say, to the use of individual legislator's offices, staff, and equipment).

Why should we care about getting the privateness and publicness of goods, rights, and owners straight? Is this simply a theoretical issue to keep scholars busy, or are there practical implications? Not surprisingly, this chapter argues that there are serious practical consequences that make definitional clarity worthwhile. First, in examining privateness and publicness of goods we slip easily into thinking that this dyad of private goods and public goods is complete when it is not. In fact, since we know that private goods are not problematic (they get produced in just the quantities we want, and efficiently too, and they are subject neither to nonprovision nor to depletion), this dyad would lead us to conclude that all of our problems arise from pure public goods. The omission of common-pool goods from the public-private dyad is dangerous because it is in precisely the overlooked but growing class of common-pool goods that almost all environmental resources fall. Second, in separating goods from property rights we can improve the match or fit between property rights and goods, improving our ability to provide and maintain common-pool goods. People could bullheadedly insist on creating fully individualized and parceled private property rights on common-pool resources and end up with management problems because they do not acknowledge the interactive features of natural production on these kinds of resources. Conversely, people could also decide, possibly for reasons of ideology or romantic nostalgia, to create common-property regimes to govern perfectly private goods that require no coordination among persons for their management. Third, and quite irksome to me, if we fail to sort out the publicness and privateness of owning entities, we risk falling into the simplistic and the sloppy habit of thinking that only individual persons can be private entities capable of owning private property and overlook the possibility that groups of individuals can be private organizations whose individual members share private rights. Definitional clarity is a prerequisite for understanding how a *group* of individuals might be a *private owner* that can share property rights and thus create a regime of *common property rights* for managing *common-pool goods*. It is a foundation on which we can begin to detect the circumstances in which common-property arrangements are appropriate, desirable, and even in some situations utterly essential to sound resource management.

Most of the permutations and combinations of resource types, property-rights types, and rights-holders theoretically exist. Surprisingly, there is very little agreement about which of these combinations and permutations are wise or efficient. There is overwhelming consensus on perhaps only two points about the appropriate combination of property rights and goods: (1) that private goods are best held as private property and (2) that private property is an inadequate arrangement for public goods and bads (that is, where we have positive or negative externalities).[5] There is also consensus, though weaker, on the inefficiencies due to principal-agent problems and rent-seeking that inevitably follow from vesting ownership in any entity other than a single individual with a central nervous system. Thus, there is considerable controversy over when it improves matters (whatever the criterion for improvement that one chooses) to vest ownership in public entities or collectivities. And we are left with a gnawing problem. What kind of property-rights arrangement *do* we design when we know that simple individual private property is inadequate, when there are externalities, and when we are concerned with pure public goods and common-pool goods? These are not problems we can ignore: human beings want public goods and common-pool goods and deserve to have them efficiently provided, and natural-resource systems on which we depend utterly are, like it or not, common-pool resources.

Because of the errors itemized above, the campaign to "privatize" ignores the nature of the goods or resources involved and confuses owners, rights, and goods with each other. By assuming that many of these resources are problematic "public goods" and therefore need "converting" into nonproblematic "private goods" (the only other class of goods they may recognize), the privatizers often imagine that they can change the nature of the good. Instead, of course, they should recognize the nature of the good as a given and recognize that what humans can manipulate are systems of rights and the identity of owning entities. Failing to recognize the nature of common-pool resources, privatizers too readily campaign on behalf of chopping up natural resource systems into environmentally inappropriate bits and pieces and of awarding rights in the bits to individuals—rather than maintaining resource systems as productive wholes and awarding rights to groups of individuals (private groups of private individuals). The danger of this fuzzy thinking—

collapsing goods, rights, and owners into a single blur, and imagining that private goods/rights/owners and public goods/rights/owners subsume the universe of possibilities—is that we have no adequate way to recognize or classify common-property regimes for common-pool goods, we misdiagnose the cause of our difficulties as the failure to force all goods to be private goods, we destroy functioning common-property regimes that already exist, and we fail to create them where they should be considered. These misunderstandings have produced tragic consequences in the handling of forests, which are themselves common-pool resources but also produce many extractable common-pool goods as well, and where inappropriate experimentation in property-rights arrangements has led to much undesirable deforestation throughout the world. The rest of this chapter applies these definitions and cautionary notes to an analysis of forests as common-pool goods (neither private goods nor pure public goods) and the common-property regimes (systems of shared private rights owned by private entities) that have been and can still be devised to manage them.

Common-Property Regimes: Problem or Solution?

Common-property regimes, used by communities to manage forests and other resources for long-term benefit, were once widespread around the globe. Some may have disappeared naturally as communities opted for other arrangements, particularly in the face of technological and economic change, but common-property regimes for forests seem in most instances to have been legislated out of existence. This happened several different ways: where common-property regimes, however elaborate and long-lasting, had never been codified, they may simply have been left out of a country's first attempt to formalize and codify property rights to forests (as in Indonesia, Brazil, and most of sub-Saharan Africa). Where common-property regimes had legal recognition, there may have been in essence a land reform that transferred all such rights to particular individuals (as in English enclosure) or to the government itself in a massive nationalization of forests, or both (as in India and Japan).

Among the many justifications usually advanced for eliminating community ownership of forests was the argument that individual or public ownership would offer enhanced efficiency in resource use and greater long-term protection of the resource. But in many instances around the world today, it is apparent that the arrangements that emerged to replace common-property regimes are ineffective in promoting sustainable resource management. Where people still live near the forests that their lives depend on, the transfer of their traditional rights into other hands does not simultaneously transfer the physical opportunity to use these resources. The people who live nearest these forests still have ample opportunity to use them, but when they lose secure property rights in the resources to others, they also lose any incentive they might have felt in the past to manage these resources for maximum long-term benefit. Now they might as well compete with each other and new users and claimants in a race to extract as much short-term benefit from the resource as possible. Thus in many instances, the transfer of property rights from traditional user groups to others eliminates incentives for monitoring and restrained use, converts owner-protectors into poachers and thus exacerbates the resource depletion it was supposedly intended to prevent. We have seen this sequence repeated wherever common-property forests have been nationalized: India and Nepal offer acute cases, but this is also true in sub-Saharan Africa and even in Meiji Japan. Thus, there is renewed interest now both in the lessons to be learned from successful common-property regimes of the past and present (see McKean, 1992a, 1992b; Netting, 1981; Berkes, 1992; Agrawal, 1994; Blomquist, 1992; Ostrom, 1986; and Thomson, 1992) and in the possibility of reviving community ownership or management as a practical remedy where appropriate.

Far from being quaint relics of a hunter-gatherer or medieval past, common-property regimes may be what we need to create for the management of common-pool resources, at least if we can identify the factors and conditions that lead to successful regimes. Sharing rights can help resource users get around problems of exclusion. They can patrol each other's use, and they can band together to patrol the entire resource system and protect it from invasion by persons outside of their group. Institutional solutions for the exclusion problem, then, begin to solve the problems of provision and maintenance. The property rights in a common-property regime can be very clearly specified, they are by definition exclusive to the coowners (members of the user group), they are secure if they receive appropriate legal support from governments, and in some settings they are fully alienable.[6]

Scholars who have designed taxonomies to point out the difference between open-access arrangements (no arrangements, rules, or property rights at all) and common property usually distinguish four "types" of property: public (state-owned), private, common, and open access (Berkes et al., 1989; Feeny et al., 1990; Bromley and Cernea, 1989; and Ostrom, 1990). Although it is extremely important to recognize that common-property regimes are not open access, this four-way taxonomy unfortunately creates the regrettable impression that common property is not private property either and does not share in the desirable attributes of private property. I think it is extremely important to point out here that common property is shared private property and should be classified just as we classify business partnerships, joint-stock corporations, and cooperatives.

Sharing private property does have its weaknesses: all arrangements of shared private property, from firms to resource cooperatives, contain internal collective-action problems because they are comprised of more than one individual owner. Just as there can be shirking and agency problems in a firm, there can be temptations inside a common-property regime to cheat on community rules. But there are productive efficiencies to be captured through team production that may be larger than losses due to shirking, making centralized or large-scale forms of production like the firm worthwhile anyway. Similarly, there may be gains from joint management of an intact resource that can outweigh losses due to cheating (or the cost of mechanisms to deter cheating) in a common-property regime (Coase, 1937; Miller, 1993).

Advantages of Common-Property Regimes

Once we understand the difference between goods and property rights (discussed above), we can understand common-property regimes as a way of privatizing the *rights* to goods without dividing the *goods* into pieces. Common property arrangements offer a way of parceling the *flow* of skimmable or harvestable "income" (the interest) from an interactive re-

Table 2.2
Stock and flow attributes of property-rights regimes

	Individual Property Rights	Common Property Rights	Public Property Rights
Rights to flow	Parceled	Parceled	Intact
Rights to stock	Parceled	Intact	Intact

source system without parceling the *stock* or the principal itself. Many natural resource systems can be far more productive when left intact than when sliced up, suggesting that they should be managed as intact wholes, or certainly in large swathes, rather than in uncoordinated bits and pieces. This is particularly true of forest ecosystems. Inherent in this basic characteristic of common property—the combination of individually parceled rights to flow with shared rights to an intact stock—lies the explanation for its appearance among human institutions. Historically, we find common-property regimes in places where a resource production system gets congested (demand is too great to tolerate continuing open access nonmanagement) so property rights in resources have to be created, but some other factor makes it impossible or undesirable to parcel the resource itself (see table 2.2).

Indivisibility

Some resources have physical traits that literally prevent parceling; the production system may simply not be amenable to physical division or demarcation. Either the resource system cannot be bounded (the high seas, the stratosphere), or the resources we care about are mobile over a large territory (air, water, fish, wildlife, plant species that propagate at great distances). Land, particularly forests, may seem much more divisible (and fenceable) at first glance than other kinds of resource systems, but in fact forests may cease to produce some of the products and benefits we want from them if we try to divide them into smaller parcels. For instance, plant species in biodiverse forests that yield fruits, nuts, gums and latex, fuelwood, timber, wildlife, and other products often depend on animal hosts for dispersal of seed and obtain protection against disease through this dispersal; these respond poorly to monoculture plantations that could be managed in smaller units. Perhaps more important and less manipulable by humans, where forests are being managed not only for products that can be taken from them but also for their value in protecting water, soil, and local climate, forests need to be managed in large units at least the size of watershed basins. Resource systems like these have to be managed in very large units. Humans have only recently acquired interest in biodiversity, but leaving natural systems unparceled and managing them in large units multiplies the biodiversity provided, sometimes exponentially, compared to managing the same acreage in separated parcels.

Uncertainty in Location of Productive Zones

In fragile environments, nature may impose great uncertainty on the productivity of any particular section of a resource system, and the location of the unproductive sections cannot easily be predicted from year to year, but the "average" or "total" productivity of the entire area may be fairly steady over the years. Management efforts focused on the entire system are not plagued with uncertainties and may therefore be quite successful. In this situation, the resource system holds still and may even have fairly obvious outer boundaries, but the productive portions of it do not hold still. This is a feature of arid forests, where nature imposes compulsory fallowing by randomly rendering certain portions of them unproductive. But it can also be true of particular production species found in rich moist forests as well, where trees used for their fruit, nuts, bark, or latex can be tapped sustainably only if they are allowed to "rest" between harvests, and thus the location of harvesting must move. In such forests, resource users may well prefer to share the entire area and decide together where to concentrate use at a particular time, rather than parceling the area into individual tracts and thereby impose the risk of total disaster on some of their members (those whose parcels turn out to be bad ones that year). Creating a common-property regime is a way of acknowledging that this risk is substantial and sharing it rather than imposing all of the risk, randomly, on some particular users each year.

Productive Efficiency via Internalizing Externalities

In many resource systems—hilly ones, for instance—uses in one zone immediately affect uses and productivity in another: deforesting the hillside ruins the water supply and downhill soil quality and may also have a negative impact on adjacent patches of surviving forest as the water table drops. If different persons own the uphill forests and the downhill fields—or, for that matter, small adjacent patches of forest and pasture—and make their decisions about resource use independently and separately, they may well cause harm to each other that requires numerous one-on-one negotiations to alleviate (Coase, 1960). An institutional alternative to this series of bilateral exchanges is to create a common-property regime to make resource-management decisions jointly, acknowledging *and internalizing* the multiple negative externalities that are implicit in resource use in this setting. People who use a common-property regime to manage their uphill forests all share ownership of the upland forests, restrain timbering to prevent soil erosion and damage to fields below, and earn more from their downhill farms than they sacrifice by not cutting as much uphill timber or fuelwood. Just as a Coaseian exchange permits people to enhance their joint efficiency by dealing directly with an externality, so joint resource management through common-property regimes may enhance efficiency by internalizing what would have been externalities in a system of individual woodlots. Common-property regimes may become desirable when more intensive resource use multiplies Coaseian considerations due to externalities between parcels. There is probably some threshold at which economies of scale in negotiating take over, and collective decision making, collective agreement on fairly restrictive use rules, and collective enforcement of those rules become easier (less time, lower transaction costs for the owners) than endless bilateral deals.

Administrative Efficiency

Even if resources are readily divisible into parcels, where nature is uniform in its treatment of different parcels so that risk and uncertainty are low, and where intensive independent use of adjacent parcels does not produce problematic externalities (perhaps this forest is on flat land, as in the margins of African savanna or sparsely vegetated dry forests), the administrative support to enforce property rights to individual parcels may not be available. The society may be too poor to support a large court system to enforce individual land titles, and even cheap fencing would be expensive by this society's standards. Creating a common-property regime here is a way of substituting collective management rules—which function as imaginary fences and informal courts internal to the user group—for what is missing. It is cheaper in these circumstances, and it is within the power of a group of resource users to create (even if they cannot create a nationwide system of courts and cannot afford barbed wire). Common-property regimes can be particularly attractive in providing administrative efficiency when resource management rules can simply be grafted onto the functions of a preexisting community organization.

In many situations, particularly where people are interested in making good use of a resource system capable of generating multiple products, more than one of these conditions applies. All around the world we have such situations: ecologically fragile uplands that make vital contributions to the livelihoods of poor people. The reasoning above would indicate that common property may be the most efficient form of property institution for such situations. We do seem to be increasingly willing to understand that nomadic pastoralism or agropastoralism based on common-property arrangements are the most productive use of arid lands that can support limited and occasional grazing and temporary cultivation but nothing else. The poor soils of the African continent, a geologic misfortune not likely to be remedied by humans,[7] may not tolerate much agricultural intensification and may need, in the long run, to be managed in large units with long fallowing periods. Indeed, the traditional common-property arrangements used for agroforestry in some areas of sub-Saharan Africa intentionally created forest "islands" as a way to improve soil for future cultivation (Cline-Cole, 1996; Fairhead and Leach, 1996). These long-term sequences of forest and fallow are a situation for which common property is very well suited.

Even in forest and other resource systems that seem eminently divisible, where risk and uncertainty are low and uniform across the resource system, where externalities seem minor or manageable through individual contracting, and where administrative support for individually owned parcels is ample, there may be reasons to maintain common property at least at some level. Natural resource systems are fundamentally interactive—forests provide watershed control, species are interdependent in ways we are often unaware of, and so on—and may well be more productive in large units than in small ones. To optimize the productivity of their own parcel, owners of individual parcels may want to guarantee that owners of adjacent parcels stick to compatible and complementary uses on their parcels, maintain wildlife habitat and vegetative cover intact, allow wildlife transit, refrain from introducing certain "problem" species, and so on. In effect, owners of individual but contiguous parcels may have an interest in mutual regulation of land use—the equivalent of zoning.[8]

To review then: private property rights in resources evolve only when demand for those resources makes the extra effort of defining and enforcing property rights worthwhile—that is, when resource use intensifies beyond some point. These may take the form of common property rights—individually owned rights to flow based on shared rights to stock—when it is impossible, undesirable, or very expensive to divide the stock (the resource base or production system) into parcels. A common-property regime consists of joint management of the resource system by its coowners and is more likely to exist when the behavior of individual resource users imposes high costs on other resource users—that is, as mutual negative externalities multiply. Vesting clear, specific, secure, and exclusive rights in private entities encourages investment and protection of resources. Vesting those rights in large enough groupings of individual resource users so that they can then coordinate their uses to match ecosystem requirements internalizes environmental externalities.

Embedded in this observation is an important theoretical proposition. That is, mutual regulation through the institutional equivalent of a common-property regime is more desirable because of its capacity to cope with multiplying externalities, as resource use *intensifies* and approaches the productive limits of the resource system. Further, since it is people who use resources, we should also find that common property becomes more desirable—not necessarily more workable but more valuable and

thus more worth trying—as population density *increases* on a given resource base. If human beings depend on extracting as much out of a resource system as the system can sustainably offer, then careful mutual
fine-tuning of their resource use becomes essential. Common-property
regimes are essentially a way to institutionalize and orchestrate this kind
of fine-tuning when resource systems are pushed to their limits.

Private property rights stimulate long-term planning, investment in the
productive quality of a resource base, and stewardship. Sharing these private property rights is a way to solve some of the externality problems
that arise from population pressure and intensification of use. Too many
observers and policymakers today now throw up their hands in despair
when they see population pressure and resource depletion, condemn common property as quaint and unworkable, and recommend privatization.
But what they mean by *privatization,* as they use the term, is either an
outright award of the entire resource system to a single individual, without regard to the political consequences of enraging all other former users
of the resource, or parcelization, rather than shared private property or
common property, which should be encompassed in the notion of
privatization.

The advocacy of "privatization," then, tends to overlook what may,
in fact, be the most appropriate form of privatization in some instances. I
would argue that common-property regimes may be the most appropriate
things to create where resource systems are under *both* environmental
and population pressure, at least where prevailing cultural values support
cooperation as a conflict-solving device. Like individual parcelization,
common property gives resource owners the incentive to husband their
resources, to make investments in resource quality, and to manage them
sustainably and thus efficiently over the long term. But unlike individual
parcelization, common property offers a way to continue limited harvesting from a threatened or vulnerable resource system while solving the
monitoring and enforcement problems posed by the need to limit that
harvesting. Sharing the ownership of the resource base is simply a way
of institutionalizing the already obvious need to make Coaseian deals to
control what are externalities for a parceled system and internalities for
a coowned intact system.

Attributes of Successful Common-Property Regimes

The findings to date from many individual case studies of successful and failed common-property regimes can be initially synthesized into a set of broad policy recommendations related to the conditions that are associated with successful common-property regimes (based on Ostrom, 1990; McKean, 1992b; and Ostrom, Gardner, and Walker, 1994).

User groups need the right, or at least no interference with their attempt, to organize   There is a stark difference between forest-user groups such as those in Switzerland and Japan that have both legal standing as property-owning entities and long-documented histories of community resource management, and indigenous peoples from Kalimantan to Irian Jaya to the Amazon, and from Zaire to India, who have practiced community resource management for decades or even centuries but have no legal protection. As soon as products from the resource system become commercially attractive, persons outside of the traditional user community become interested in acquiring legal rights to the resource. If the traditional users have those legal rights in the first place, then they essentially have the commercial opportunities that their resources create. In Papua New Guinea, for instance, where traditional community forest rights are legally valid, portable sawmills used by villagers turn out to be more economically efficient overall, and to bring more wealth into the village, than timbering by multinational corporations. Where local communities' resource claims go unrecognized by national governments, the best they can then hope for is that higher layers of government will overlook them rather than oppose them. The farming villages of Andhra Pradesh that use an open-field system to manage planting, harvesting, grazing, and irrigation do so successfully only because and as long as the state and national governments ignore them (Wade, 1992).

The boundaries of the resource must be clear   It is easier to identify and define both the natural physical boundaries for some resources (forests are an obvious example) and the legal boundaries for a particular community's land than it is to define boundaries for, say, a highly mobile species of fish in the high seas. Once defined, these boundaries can then be patrolled by community guards. Clearly marked or even well understood boundaries can be an inexpensive substitute for fencing. Indeed, fencing may be an effective barrier against some animals but not against human beings, who can climb over most fences and, in any case, usually acquire wire clippers and saws at the same time they get hold of fencing material. Rather, the social function of fencing, one that can be performed equally well by unambiguous demarcation of property lines, is that it offers impartial notification of boundaries. Thus, those who invade others' territory know they are doing it, and those who are invaded can prove readily that they have been invaded. Fencing eliminates innocent error and ignorance as excuses for trespass and theft.

The criteria for membership in the group of eligible users of the resource must also be clear   The user group has to share solid internal agreement over who its members are, and it is probably best if eligibility criteria for membership in this group do not allow the number of eligible users to expand rapidly. Many Swiss villages limit eligibility to persons who live in the village *and* purchase shares in the alp, so that new residents must find shares to buy, and shareowners who leave the village find it in their interest to sell their shares because they are unable to exercise their village rights from elsewhere (Netting, 1981; Glaser, 1987). Thus, the size of the eligible user group remains stable over time. Japanese villages would usually confer eligibility and shares of harvest on households rather than individuals and were also likely to limit membership to long-established "main" households rather than "branch" households. These practices ensured that no special advantages went to large households, those that split, or new arrivals. Not only did this rule limit the number of eligible users and the burden on the commons, but it also discouraged population growth (McKean, 1992a). Communities elsewhere may be less strict—at their peril—about defining eligibility for membership in the user group. Vondal describes an Indonesian village whose communal resources are under stress in part because the community opens membership in the user group not just to all village residents but also to all kin in neighboring villages (in McCay and Acheson, 1987). Thus, this user group has expanded rapidly, without any consideration yet for matching its size or its aggregate demand for resources to the capacity of the resource system.

Users must have the right to modify their use rules over time   Inflexible rules are brittle and thus fragile and can jeopardize an otherwise well-organized common-property regime. In a magnanimous but ill-considered attempt to extend legal recognition to common-property regimes over forest and pasture land in the Punjab, the British decided to codify all of the rules of resource use in different systems. The undesirable consequence was to freeze in place use rules that really needed to remain flexible (Kaul, 1995). The resource users are the first to detect evidence of resource deterioration and resource recovery and so need to be able to adjust rules to ecological changes and new economic opportunities. If the commons displays signs of distress, the village might alter the rules so as to reduce or even eliminate the incentive for each family to cut all that it can when allowed entry into the commons. The village might choose to lengthen the period of closure on a forest that is being degraded. Or it could alter distribution rules from allowing each family to keep what one able-bodied adult can bring out of the commons in one day during entry season, to aggregating the cut from each family, dividing it into equal amounts, and reassigning bundles of harvest to each household by lottery. Japanese villages that have retained full title to their common lands are not only free to adjust regular use rules as they see fit but are also free to take advantage of attractive commercial opportunities. They may hire loggers to clear 1/50 of the mountain each year for 50 years. They may "manage" the forest for commercially valuable bamboo or fruit trees. Or villages may lease surface rights to hotels and ski resorts. They are even free to sell off the commons, by unanimous vote, if they want to reap the capital gains on appreciated land values.

Use rules must correspond to what the system can tolerate and should be environmentally conservative to provide a margin for error   Successful user groups appear to prefer environmentally conservative use, possibly to give themselves a margin to invade during emergencies. Japanese villagers in the Mt. Fuji area knowingly overused their forest commons during the depression of the 1930s (removing more fodder for packhorses and more wood for charcoal than they should have) but also knew that they—and the commons itself—could afford this in a temporary emergency of that kind precisely because they were intentionally conservative in their use during good times. The commons was both an essential part of everyday living and a backup system maintained in reserve. When forestry scientists told Nepali villagers that their forest could easily tolerate the extraction of both leaf litter and kindling, the villagers rejected this advice and opted instead to ban the cutting of fuelwood altogether because they feared that allowing any cutting of wood would threaten the total population of deciduous trees and thus could reduce the supply of the leaf litter they used as fodder and fertilizer (Arnold and Campbell, 1986).

Use rules need to be clear and easily enforceable (so that no one need be confused about whether an infraction has occurred)   Common-property regimes frequently establish quantitative limits on amounts of different products that an individual user may extract from various zones of the commons, but this means that a suspected infraction involves much measurement, weighing, and discussion between resource user and guard about whether this limit applies to that species or another one, whether this kindling was collected from one zone or two, whether these branches are of too wide a diameter or not, and so on. Sometimes other kinds of rules can be simpler to understand and enforce. Restrictions on the equipment a user takes into the forest may be just as effective in restraining harvesting and also be simpler to enforce. Having too large a saw or a pack animal rather than a backpack might then be an infraction even before one begins to cut. Opening and closing dates are similar: being in the forest during the off season is simply unacceptable, whatever the excuse. Clear enforceable rules make life easier for resource users and for monitors representing the user group and reduce misunderstandings and conflict.

Infractions of use rules must be monitored and punished   Obviously, rules work only when they are enforced. Agrawal (1992) found that communities in Uttar Pradesh differ widely in the extent to which they devote village resources to enforcement, particularly hiring guards or assigning villagers to guard duty by some rotational scheme. The communities with healthy common forests were those that recycled the fines and penalties they collected into providing for their guards. The communities with degraded forests were those that had fewer guards, enforced the rules less, collected much less in fines, and put the fines into a general village budget rather than into the enforcement mechanism. There is also evidence that penalties need not be draconian: graduated penalties, mild for first offenses and severe only for repeated infractions, are adequate (McKean, 1992b; Ostrom, 1990).

Distribution of decision-making rights and use rights to coowners of the commons need not be egalitarian but must be viewed as "fair" (one in which the ratio of individual benefit to individual cost falls within a range they see as acceptable)   It comes as a surprise to observers who have romanticized the commons that common-property regimes do not always serve to equalize income within the user group. Communities vary enormously in how equally or unequally they distribute the products of the commons to eligible users. Decision-making rights tend to be egalitarian in the formal sense (one user household, one vote), although richer households may actually have additional social influence on decisions. Entitlement to products of the commons varies to a surprising extent (McKean, 1992b).

In some communities, especially in India, the commons do turn out to be a welfare system for the poor: the wealthy members of the community may be entitled to use the commons but do not bother to exercise that right because of the high opportunity cost of their labor, leaving *de facto* access to poorer members—those willing to invest their labor in collecting products from the commons.

In other communities, including most long-lived common-property regimes (Switzerland, Japan, and virtually all regimes governing grazing and irrigation), products of the commons are distributed to families in the same proportions as their private assets off of the commons. If any subgroup feels cheated—denied "adequate" access or a "fair" share—compared to another subgroup, the angry subgroup becomes unwilling to participate in decision making, unwilling to invest in maintaining or protecting the commons, and motivated to vandalize the commons. An important key to the cohesiveness of farmer-managed (as opposed to government-organized) irrigation systems is the power of tailenders to withhold their labor from maintenance of canals, channels, and sluicegates when they feel that headenders are taking too much water. Successful irrigation systems have very well-calibrated mechanisms to distribute water in the same proportions as the labor required of coowners (Tang, 1992). Rules that award more benefits to those who invest more, and no benefits to those unwilling to invest, seem to have the best chance of winning the allegiance of both rich and poor.

Inexpensive and rapid methods are needed for resolving minor conflicts   Successful common-property regimes assume that there will often be small disagreements among users and provide regular opportunities for these disagreements to be aired and rules clarified or adjusted if necessary. Swiss commoners make Sunday church outings the regular occasion for discussing problems and collecting levies. Japanese villagers are so organized (it is not unusual to find more committees than households in a village) that they have constant opportunities to air grievances. Most conflicts can be resolved at a low level because persons with multilayered social relationships can usually design a satisfactory compromise.

Institutions for managing very large systems need to be layered with considerable devolution of authority to small components to give them flexibility and some control over their fate   Some forests, grazing areas, and irrigation systems may have to be managed in very large units, but at the same time the persons living near each patch or segment of the resource system need to have substantial and secure rights in the system to have the incentive to protect the portion near them. A large resource system may be used by many different communities, some in frequent contact with each other and some not. The need to manage a large resource system as a unit would seem to contradict the need to give each of that resource system's user communities some independence. Nesting different user groups in a pyramidal organization appears to be one way to resolve this contradiction, providing simultaneously for independence and coordination. The most successful examples of complex nesting come from irrigation systems serving thousands of people at a time (Ostrom, 1990, 1992), but there are also examples of nested arrangements for common forests, in Japan (McKean, 1996).

It must be recognized that some common-property regimes falter and that other sorts of institutional arrangements can also work effectively. But it would be a grave mistake to dismiss common-property regimes as relics of the past, intrinsically unworkable, or incompatible with contemporary society. The theoretical arguments above indicate that in some circumstances common-property regimes may be quite suitable, and, in fact, many documented cases show resource users who themselves have crafted institutions consistent with our findings above. But there are still many gaps in our knowledge and information about the effects of diverse institutions on forest conditions. Before we destroy or create institutions willy-nilly, we need much continued effort to enlarge the body of information we draw on in the effort to reduce rates of deforestation and loss of biodiversity around the world.

Although we are a long way from certainty about what makes successful common-property regimes work, I would be willing to offer the following propositions:

• *Sociocultural support*   Common-property regimes will work better where the community of users is already accustomed to negotiating and cooperating with each other on other problems than where there are numerous existing conflicts and no indication of a willingness to compromise.

• *Institutional overlap*   Reviving recently weakened institutions, where the habits and techniques of negotiation and compromise are still in evidence, will be easier than trying to invent wholly new institutions among people who have never worked together before.

• *Administrative support*   Reviving or creating common-property regimes where local and national governments are hostile is almost impossible. There is little point in trying unless local and national elites, or significant portions of them, are sympathetic to the attempt. This kind of support means legal recognition to strengthen the security and enforceability of common property rights.

• *Financial support*   Apart from limited help with local start-up costs, financial support to local common-property regimes is probably *un*desirable because it might well undermine local cooperation. If an institutional form is being adopted because it is efficient, it should pay for itself (by definition) and not require subsidy.

• *Conflict reduction*   Where the size of productive management units permits a certain degree of segmentation or parceling of the resource, it is probably preferable to create nonoverlapping commons for different communities or nested arrangements rather than to have several communities sharing a single huge commons. It is probably best for the communities involved to make this choice rather than to have an outsider insist on splitting the resource system into several separate commons.

Common-property regimes are being promoted at long last in a number of resource-poor developing countries as a way of restoring degraded forests and building up a community resource base (note the instance of Joint Forest Management programs in India and the revival of community forestry in Nepal). I argue here that common property may be more appropriate than individual property when externalities among parcels of land multiply due to intensive use and high population pressure. It is crucial, then, not to eliminate common-property arrangements where they survive but rather to view common property as a legitimate and very suitable variety of private property in some circumstances when conducting property-rights reform and to pay careful attention to the nature of the resources in question (are they common-pool goods?) before tampering with property rights to those resources.

Acknowledgments

An earlier version of this paper was presented at the International Conference on Chinese Rural Collectives and Voluntary Organizations: Between State Organization and Private Interest, Sinological Institute, University of Leiden, 9–13 January 1995, and is based on portions that I contributed to a paper coauthored with Elinor Ostrom, "Common Property Regimes in the Forest: Just a Relic from the Past?," *Unasylva* 46(180) (January 1995): 3–15. I would like to thank David Feeny, Clark Gibson, Elinor

Ostrom, and three anonymous reviewers for thoughtful suggestions. I remain responsible for any errors.

Notes

1. I prefer to avoid the often used term *common-property resources* because it conflates property (a social institution) with resources (a part of the natural world). I will also avoid using the acronym *CPR* in the text that follows, since that could easily stand for any of the three terms (*common-property resources, common-pool resources,* or *common-property regimes*—not to mention *cardiopulmonary resuscitation*).

2. Garrett Hardin's (1968) classic essay on the tragedy of the commons points out the hazards of open access, without stating clearly that the problem was the lack of a property-rights or management regime (the openness of access), not the sharing of use (common use). Hardin (1994) has taken steps to rectify this oversight in more recent work that distinguishes between the unmanaged (unowned) commons subject to tragedy and the *managed* (owned) commons where property rights may be able to prevent misuse of the resource.

3. The nature of a good can change with technology. Thus, TV broadcasts from satellites are pure public goods when the satellite signals are unscrambled. The advent of scramblers, cable services, and purchasable descrambler boxes converts TV broadcasts into excludable and nonsubtractable goods (thus toll goods or club goods). The advent of cheap illegal descramblers converts TV broadcasts back into nearly public goods again. But at any particular technological moment, the nature of a good is indeed a given.

4. This definition obviously does not include all governments. Many autocratic governments neither intend nor accomplish the representation of the general public and would be better described as private government.

5. American economists (North, 1990; North and Thomas, 1973; Demsetz, 1967; Alchian and Demsetz, 1973; Anderson and Hill, 1977; Libecap, 1989; Johnson and Libecap, 1982) have argued persuasively that property rights emerge in response to conflict over resource use and conflicting claims over resources, and that well-defined property rights help to promote more efficient use of resources and more responsible long-term care of the resource base. This evolution is probable but not guaranteed, and conflict over resource use can simply continue without efficiency-enhancing evolution of clearer property rights. Tai-Shuenn Yang (1987) argued that retention of residual imperial prerogatives over all resources in China made all property rights that did evolve there merely temporary and insecure and inhibited economic growth in China for two millennia. Peter Perdue (1994) disputes this view, however.

6. Some Swiss alpine common-property regimes, some Japanese agricultural and forest common-property regimes, and all Japanese fishing cooperatives permit trading in shares (the individually parceled rights to flow or income), and all have

mechanisms by which the entire common-property user group may actually sell its assets (the shared rights to stock or capital assets of the user group or corporation) (Netting, 1981; Glaser, 1987; McKean, 1992a).

7. The African continent, having been the one from which other continental plates split off, was not fortunate enough to have been crashed *into* by other plates. It is this collision of plates that produces a gigantic upwelling of old sea floors into new mountain ranges, and it is such mountain ranges that over geologic time erode into the rich alluvial plains of the world's breadbasket regions. The mountains formed (as in East Africa) when a plate slides across areas of volcanic eruption consist of molten lava with no organic enrichment, and although they too erode and contribute to topsoil, it is of much lower agricultural value (David Campbell, Department of Geology, Michigan State University, personal communication, 28 June 1995).

8. In fact, zoning and urban planning are actually the creating of common or shared property rights in choices over land use and the vesting of those rights in the citizens of a municipality. Just as zoning would be an absurdly unnecessary effort in a frontier area where population density is low but increasingly desirable—to control externalities—in more densely populated areas, so common property becomes *more* desirable, not less, with more intense resource use.

References

Agrawal, Arun. 1992. "Risks, Resources, and Politics: Studies in Institutions and Resource Use from India." Ph.D. diss., Duke University.

———. 1994. "Rules, Rule Making and Rule Breaking: Examining the Fit between Rule Systems and Resource Use." In *Rules, Games, and Common-Pool Resources,* ed. Elinor Ostrom, Roy Gardner, and James Walker, 267–82. Ann Arbor: University of Michigan Press.

Alchian, Armen A., and Harold Demsetz. 1973. "The Property Rights Paradigm." *Journal of Economic History* 33(1) (March): 16–27.

Anderson, Terry L., and P. J. Hill. 1977. "From Free Grass to Fences: Transforming the Commons of the American West." In *Managing the Commons,* ed. Garrett Hardin and John Baden, 200–15. San Francisco: Freeman.

Arnold, J. E. Michael, and J. Gabriel Campbell. 1986. "Collective Management of Hill Forests in Nepal: The Community Forestry Development Project." In *Proceedings of the Conference on Common Property Resource Management: April 21–26, 1985 (Annapolis, Maryland),* ed. Daniel W. Bromley et al., 425–54. Washington, DC: National Academy Press.

Berkes, Fikret. 1992. "Success and Failure in Marine Coastal Fisheries of Turkey." In *Making the Commons Work: Theory, Practice, and Policy,* ed. Daniel W. Bromley et al., 161–82. San Francisco: ICS Press.

Berkes, Fikret, David Feeny, Bonnie J. McCay, and James M. Acheson. 1989. "The Benefits of the Commons." *Nature* 340 (July): 91–93.

Blomquist, William. 1992. *Dividing the Waters: Governing Groundwater in Southern California.* San Francisco: ICS Press.

Bromley, Daniel W., and Michael M. Cernea. 1989. "The Management of Common Property Natural Resources: Some Conceptual and Operational Fallacies." World Bank Discussion Papers No. 57. Washington, DC: World Bank.

Bromley, Daniel W., David Feeny, Margaret A. McKean, Pauline Peters, Jere Gilles, Ronald Oakerson, C. Ford Runge, and James Thomson, eds. 1992. *Making the Commons Work: Theory, Practice, and Policy.* San Francisco: ICS Press.

Cline-Cole, Reginald. 1996. "Dryland Forestry: Manufacturing Forests and Farming Trees in Nigeria." In *The Lie of the Land: Challenging Received Wisdom on the African Environment,* ed. Melissa Leach and Robin Mearns, 122–39. London: Currey and Heinemann.

Coase, Ronald. 1937. "The Nature of the Firm." *Economica* 4(16) (November): 386–405.

———. 1960. "The Problem of Social Cost." *Journal of Law and Economics* 3: 1–44.

De Alessi, Louis. 1980. "The Economics of Property Rights: A Review of the Evidence." *Research in Law and Economics* 2(1): 1–47.

———. 1982. "On the Nature and Consequences of Private and Public Enterprises." *Minnesota Law Review* 67(1) (October): 191–209.

Demsetz, Harold. 1967. "Toward a Theory of Property Rights." *American Economic Review* 57 (May): 347–59.

Fairhead, James, and Melissa Leach. 1996. "Rethinking the Forest-Savanna Mosaic: Colonial Science and Its Relics in West Africa." In *The Lie of the Land: Challenging Received Wisdom on the African Environment,* ed. Melissa Leach and Robin Mearns, 105–21. London: Currey and Heinemann.

Feeny, David, Fikret Berkes, Bonnie J. McCay, and James M. Acheson. 1990. "The Tragedy of the Commons: Twenty-two Years Later." *Human Ecology* 18(1): 1–19.

Glaser, Christina. 1987. "Common Property Regimes in Swiss Alpine Meadows." Paper presented at the Conference on Comparative Institutional Analysis, Inter-University Center of Postgraduate Studies, Dubrovnik, Yugoslavia, 19–23 October.

Gordon, H. Scott. 1954. "The Economic Theory of a Common Property Resource: The Fishery." *Journal of Political Economy* 62 (April): 124–42.

Hardin, Garrett. 1968. "The Tragedy of the Commons." *Science* 162: 1,243–48.

———. 1994. "The Tragedy of the Unmanaged Commons." *Trends in Ecology and Evolution* 9: 199.

Johnson, R. N., and Gary D. Libecap. 1982. "Contracting Problems and Regulation: The Case of the Fishery." *American Economic Review* 72(5) (December): 1,005–23.

Kaul, Minoti Chakravarty. 1995. Personal communication.

Libecap, Gary D. 1989. *Contracting for Property Rights.* New York: Cambridge University Press.

Locke, John. 1965. "On Property." In *The Second Treatise* (An Essay Concerning the True Original, Extent, and End of Civil Government), *Two Treatises of Government* (any edition), ch. 5, paras. 25–51, pp. 327–44. New York: Mentor.

McCay, Bonnie J., and James M. Acheson. 1987. *The Question of the Commons: The Culture and Ecology of Communal Resources.* Tucson: University of Arizona Press.

McKean, Margaret A. 1992a. "Management of Traditional Common Lands (*Iriaichi*) in Japan." In *Making the Commons Work: Theory, Practice, and Policy,* ed. Daniel W. Bromley et al., 63–98. San Francisco: ICS Press.

———. 1992b. "Success on the Commons: A Comparative Examination of Institutions for Common Property Resource Management." *Journal of Theoretical Politics* 4(3): 247–82.

———. 1996. "Common Property Regimes as a Solution to Problems of Scale and Linkage." In *Rights to Nature,* ed. Susan Hanna, Carl Folke, and Karl-Göran Mäler, 223–43. Washington, DC: Island Press.

Miller, Gary J. 1993. *Managerial Dilemmas: The Political Economy of Hierarchy.* Cambridge: Cambridge University Press.

Netting, Robert McC. 1981. *Balancing on an Alp.* New York: Cambridge University Press.

North, Douglass C. 1990. *Institutions, Institutional Change and Economic Performance.* New York: Cambridge University Press.

North, Douglass C., and R. P. Thomas. 1973. *The Rise of the Western World: A New Economic History.* Cambridge: Cambridge University Press.

Ostrom, Elinor. 1986. "An Agenda for the Study of Institutions." *Public Choice* 48: 3–25.

———. 1990. *Governing the Commons: The Evolution of Institutions for Collective Action.* New York: Cambridge University Press.

———. 1992. *Crafting Institutions for Self-Governing Irrigation Systems.* San Francisco: ICS Press.

Ostrom, Elinor, Roy Gardner, and James Walker. 1994. *Rules, Games, and Common-Pool Resources.* Ann Arbor: University of Michigan Press.

Perdue, Peter. 1994. "Property Rights on Imperial China's Frontiers." Paper presented at the Social Science Research Council's Committee on Global Environmental Change, Seminar on Landed Property Rights, Stowe, Vermont, 17–21 August.

Runge, Carlisle Ford. 1981. "Common Property Externalities: Isolation, Assurance and Resource Depletion in a Traditional Grazing Context." *American Journal of Agricultural Economics* 63: 595–606.

————. 1984. "Strategic Interdependence in Models of Property Rights." *American Journal of Agricultural Economics* 66: 807–13.

————. 1992. "Common Property and Collective Action in Economic Development." In *Making the Commons Work: Theory, Practice, and Policy,* ed. Daniel W. Bromley et al., 17–39. San Francisco: ICS Press.

Schlager, Edella, and Elinor Ostrom. 1992. "Property-Rights Regimes and Natural Resources: A Conceptual Analysis." *Land Economics* 68(3) (August): 249–62.

Schlager, Edella, and Elinor Ostrom. 1993. "Property-Rights Regimes and Coastal Fisheries: An Empirical Analysis." In *The Political Economy of Customs and Culture: Informal Solutions to the Commons Problem,* ed. T. L. Anderson and R. T. Simmons, 13–41. Lanham, MD: Rowman & Littlefield.

Scott, A. D. 1955. "The Fishery: The Objectives of Sole Ownership." *Journal of Political Economy* 63 (April): 116–24.

Tang, Shui Yan. 1992. *Institutions and Collective Action: Self-Governance in Irrigation.* San Francisco: ICS Press.

Thomson, James T. 1992. *A Framework for Analyzing Institutional Incentives in Community Forestry.* Rome: Food and Agriculture Organization of the United Nations, Forestry Department, Via delle Terme di Caracalla.

Wade, Robert. 1992. "Common Property Resource Management in South Indian Villages." In *Making the Commons Work: Theory, Practice, and Policy,* ed. Daniel W. Bromley et al., 207–28. San Francisco: ICS Press.

Yang, Tai-Shuenn. 1987. "Property Rights and Constitutional Order in Imperial China." Ph.D. diss., Indiana University, Bloomington.

3

Small Is Beautiful, but Is Larger Better? Forest-Management Institutions in the Kumaon Himalaya, India

Arun Agrawal

Introduction

An increasing number of scholars, development practitioners, and environmental activists today forward microinstitutional solutions as the remedy for renewable-resource scarcities. They have thus helped to shift attention away from market- or state-oriented policies as the only two alternatives to achieve development or environmental conservation (Anderson and Grove, 1987; Ostrom, 1990; Ostrom, Schroeder, and Wynne, 1993). The fresh claims on behalf of the local (Chambers, 1983; Korten, 1986; Uphoff, 1986), the indigenous (Cultural Survival, 1993; Denslow and Padoch, 1988; Richards, 1985), and the "little community" (Hecht and Cockburn, 1990; Scott, 1976; Wade, 1994) represent a long overdue move.[1]

The growing focus on community institutions and indigenous voices recognizes that national and international environmental trends are the aggregate consequence of the possibly independent concrete actions of millions of users. It accepts the rupture between the interests of local populations and those of national governments and international institutions. After all, advocates of global conservation or national development may alike encroach on the rights and capacities of local users of natural resources (Redford and Sanderson, 1992; Agrawal, 1992). But even more appropriately, the focus on the local marks a shift from the preoccupation with centralized, overarching, and overreaching solutions of the past decades that have failed to reverse and may indeed have contributed to environmental problems and attendant social tensions.[2] Existing state policies on development and conservation are increasingly seen to inflict violence at multiple levels on everyday relations of existence and livelihood in rural areas.[3]

The attention to local spaces and communities, thus, forms a critical move in the conversation on development and conservation. The ensuing study builds on the insights in this literature by interrogating the relationship between group size and successful achievement of collective action. Contrary to a large literature in the social sciences, I question the presumption that smaller groups are more successful than larger groups.

The study analyzes village forest councils in Almora district in the Indian Middle Himalaya. These community-level councils help residents utilize and protect forest resources in accordance with rules they themselves craft and help to enforce. To meet the objectives of this chapter, I first briefly describe the process behind the birth of the forest councils. I then examine the interactions between the interests of the British colonial state and the actions of local populations and how these led to outcomes that incorporated the interests of village populations.

The sketch of the birth of the forest councils in the region sets the stage for seeking the solution to a puzzling finding of the research: councils with a larger membership find it easier to organize successfully for collective action, and the smaller councils face difficulties in organizing successfully.[4] An enormous literature in the social sciences, inspired by the seminal work of Mancur Olson, has investigated why smaller groups are more successful in organizing collective action. The analysis seems convincing. Rational individuals, acting in their self-interest, are unlikely to act in ways that would facilitate the provision of collective goods for a group, even if all group members share the same interest. Hammering this insight home, Olson showed how smaller groups are better able to overcome the problem of collective action in comparison to larger groups. Since his work, analysts have underscored his conclusions using game theory and metaphors such as "the tragedy of the commons" (see Ostrom, 1990, for an analysis).

The findings reported in this chapter, however, undermine conventional wisdom. Building on the empirical observation that smaller forest councils find it more difficult to organize successful collective action, the chapter discusses some significant theoretical reasons why larger groups might be more successful. After describing the basic characteristics of the villages where I conducted research, I first attempt a local explanation of the success of the larger councils. I then elaborate the analysis to provide a more generally applicable explanation. In examining the relationship between group size and collective action, this study makes two important departures. Much writing on collective action focuses on the internal dynamics of a group. In contrast, this chapter looks at the external dynamics: the relations of a group with other groups. Second, it draws a distinction between mobilizing a group for collective action and success in meeting the objectives of collective action. Using these two distinctions, it constructs an argument about why larger groups may be more successful than smaller ones.

The Forest Councils of Kumaon

A multiplicity of institutional forms occupies the terrain of resource management in Almora. Three distinct regimes can be identified: (1) reserved forests controlled by the Forest Department, (2) civil forests managed by the Revenue Department, and (3) community forests managed by the forest councils. The activities of the forest councils are the focus of investigation.

I trace the history of the forest councils to the activities of the colonial British state in the mid-nineteenth century. From this period onward, the British government made a number of inroads to curtail progressively the area of forests under the control of villagers (Guha, 1990, 44–45). Between 1910 and 1917 alone, the government transferred an additional 2,500 square kilometers of forests to the Imperial Forest Department. At the same time it also enacted elaborate new rules specifying strict restrictions on lopping and grazing rights, reduced rights to nontimber forest products, prohibited the extension of cultivation, sought to regulate the use of fire that villagers believed to result in higher grass production, increased the labor extracted from the villagers, and strengthened the number of official forest guards (Pant, 1922).

The new rules stirred villagers into widespread protest. They simply refused to accept the rules or the fundamental assumption undergirding them—the state's monopoly over all natural resources it deemed significant. The best efforts of government officials failed to convince the villagers that the forests belonged to the government (Ballabh and Singh, 1988). The government had hoped that the residents of the hills "would gradually become accustomed to the rules," but "the hill man proved impatient of control" (KFGC, 1922, 2). The incessant, often violent, protests forced the government to appoint the Kumaon Forest Grievances Committee to look into the local disaffection. The Committee examined over 5,000 witnesses from all parts of Kumaon in 1921 to make more than 30 recommendations. On the basis of these recommendations, the government passed the Forest Council Rules of 1931. These rules empowered village communities to create forest councils and bring under their own control forest lands that were managed by the revenue department as Class I and Civil Forests.[5]

Nearly 3,000 forest councils today formally control about 30 percent of the hill forests in Kumaon. Of these, close to 1,700 exist in Almora alone (Agrawal, 1995, 51). The broad parameters that define the management practices of these institutions are laid down in the Forest Panchayat Rules. More specific day-to-day management of community forests is the result of local action. Rural residents meet frequently, discuss the specific rules that will govern withdrawal of benefits from forests, and create monitoring, sanctioning, and arbitration devices to resolve the vast majority of management questions at the local level. They elect their leaders from within the community, select guards to enforce rules, fine rule breakers, manage finances, and often deploy earnings for the benefit of the community.[6]

This abbreviated history of the emergence of the forest councils in Kumaon resonates with some critical issues in the social sciences. In contrast to much writing on local communities and peasants that treats its subjects as unwitting victims of a power-hungry centralizing state, it shows that in the Kumaon hills, villagers significantly influenced government policies to make them reflect local needs for forests. They organized, resisted new state policies, and gained a measure of success in wresting back some control over forests. This is not to say that state actors do not seek greater control. Rather, it is simply to underline that although macro-level initiatives can determine micro-level outcomes, the contours of such initiatives and the processes through which their outcomes unfold are unavoidably shaped by social action at the micro level.

Resources of the Councils

The most significant products villagers harvest from their forests are fodder, fuelwood, animal bedding, organic manure, and construction timber. Figure 3.1 outlines the importance of forests in the hill agricultural and subsistence economy by tracing the links between forest products, and the kinds of needs such products fill. Forests are the cornerstone of subsistence in the hills, contributing critical inputs to each element of the subsistence economy—the household, agricultural fields, and livestock rearing. In addition, council forests containing chir pine (*Pinus roxburghii*) also yield resin for turpentine, a commercially valuable product.

Subsistence products from the community forests are usually available to all residents of the villages in which the forest councils are located. The cash revenues from the distribution of the forest products are used to monitor and guard the resource and to meet operational expenses of the councils. In some cases, councils have also had sufficient surpluses

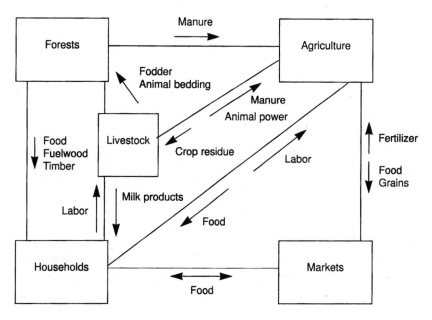

Figure 3.1
Forests in the hill subsistence economy

to create communal goods for their villages such as school buildings or common utensils that are used to cook food for the community during festive celebrations such as marriages or religious festivals.

Key Actors

The forest councils are embedded in a web of social and administrative relationships. These relationships presume the patterns of influence laid down in the Forest Council Rules of 1931, as amended in 1976 (see table 3.1). While the Rules provide for support to the councils from the Revenue and the Forest Departments to facilitate rule enforcement and the maintenance of vegetation in the forests, it grants them only limited authority to enforce rules. Indeed, over the last several decades, the modifications in the Rules and the manner of their application have greatly reduced the independence of the villagers. In the quotidian interactions of different actors that influence the performance of the councils, higher-level government officials, especially those in the Revenue Department, have emerged as pivotal in the success of the councils. That they were assigned supervisory and enforcement powers played a crucial role in this process.

As table 3.1 shows, the powers of the councils, especially their enforcement authority, suffered a substantial decline in 1976. The overall framework of rules within which they could operate became far stricter. In addition, new restrictions on day-to-day activities meant that they could fine rule breakers only with the consent of the person involved or once permission was secured from higher-level government officials. For major disputes they were required to move the judiciary or rely on aid from the officials of the Revenue Department.

As a result, those forest councils that have few resources at their command have been plagued by rule infractions. Their elected officials, lacking independent means to pursue court cases and the requisite influence to move the officials of the Revenue Department, have often been helpless to enforce the rules they created. Asked in a meeting to list the four most important problems facing their councils, 30 heads of councils listed problems related to inadequate supervision and local rule breaking and monitoring 68 percent of the time. In contrast, problems related to low cash incomes of the councils were mentioned only 32 percent of the time.[7]

At the same time, the officials of the Revenue Department who are supposed to help the councils must perform a host of other duties, including the maintenance of law and order, collection of taxes, and administration of development projects. Most government officials consider these duties to have a greater priority over tasks related to forest councils. For many councils inadequate levels of enforcement and limited local resources are a major problem.

The Case Studies

Data on nine forest councils form the basis for the ensuing discussion.[8] Five of these councils are located in the Dhauladevi Development Block of Almora District, near the historic religious site of Jageshwar.[9] The second set of four councils is drawn from the Lohaghat development block of Pithoragarh district. All the nine councils range in elevation from 1,100 to 2,000 meters; their forests lie between 1,400 and 2,100 m. They are all close to motorable roads and thus more or less equally exposed to market forces (see table 3.2). Forest resources are scarce for the residents of all the villages that have the councils, and villagers compete for subsistence benefits from forests. Many of the residents in the nearby villages, who do not have their own council forests, depend either on scattered plots of forests owned by the Forest Department or on the forests of their neighbors.

Although the selected forest councils and their settlements are situated within the same ecological and administrative divisions and face similar levels of market pressures, they differ in size, organization, age, and resource endowments. Table 3.2 presents some basic features of the selected councils. The first five are the councils from Almora district, and the latter four are from Pithoragarh. We can say that six of the councils are small (in number of households): Pokhri, Tangnua, and Kana in Almora, and Lada, Kadwal, and Jogabasan in Pithoragarh. None of them have more than 30 households. Kana and Lada are a little larger within the group of small councils. Kotuli and Bhagartola are relatively large, as is Goom. The same points about size can be made as far as the area of the council

Table 3.1
Changes in the Van Panchayat Act between 1931 and 1976

Subject	1931	1976
Formation or dissolution	1. Two or more residents can propose the formation of the van panchayat for a village. 2. The Deputy Commissioner can dissolve a panchayat in case of repeated mismanagement or rule infractions.	Rule 2 remains the same: *Modifications:* 1. One-third of the villagers must propose the formation of the van panchayat.
Membership	1. At least three, and at most nine, members are elected to the van panchayat by villagers. 2. Panches select their leader as Sarpanch. 3. Panches can force the resignation of individual members by a majority. The empty position can be filled from among rightholders by a majority decision of the panches. 4. All village residents and others who possess rights in the forest can be rightholders in the panchayat forest.	Rules 2, 3, and 4 remain the same. *Modifications:* 1. Five to nine members to be elected to the van panchayat. 2. The Deputy Commissioner can nominate one member to the panchayat. 3. The Sarpanch can be removed from office by one-third of the members, provided this step is approved by two-thirds of the members in a subsequent meeting.
Rules regarding resin extraction	1. The Forest Department is responsible for harvesting resin from chir pine trees. 2. Profits are shared between the Forest Department and the panchayat in proportions to be determined by the Forest Conservator. 3. Panchayat can harvest resin in accordance with rules laid down by the Forest Department. The resin can be sold to either the Forest Department or registered buyers.	Rules 1, 3, and 4 remain the same. *Further Restrictions:* a. See modifications c, d, and e under "Allocation of Income."

4. Panchayat members can harvest resin for domestic use.

Rules laid down by government

1. Panchayat forest land cannot be sold, mortgaged, or subdivided.

2. The products and proceeds from the sale of products of the panchayat forest are to be used for the benefit of the community.

3. The panchayat is to protect the forest and its trees (but with no explicit restriction on commercial sale of trees or timber).

4. The panchayat is to prevent villagers from cultivating the panchayat forest land.

5. The panchayat is to demarcate the forest area.

6. The panchayat is to maintain minutes of meetings and records of accounts and make decisions in regular meetings.

7. The panchayat is to follow the instructions of higher revenue officials.

8. The quorum requires two-thirds of the members of the committee to be present.

9. All decisions are to be made by simple majority.

Rules 1, 2, 4, 5, 6, 7, and 8 remain the same.
Further Restrictions:

a. All decisions of the panchayat are to be made by two-thirds vote.

b. The panchayat is to meet at least once every three months; proceedings of the meeting are to be recorded and a copy submitted to the deputy commissioner.

c. All extraction of timber beyond one tree requires permission from the Deputy Commissioner, Divisional Forest Officer (DFO), and the Conservator of Forests (CF). Any sales of forest produce must be in accordance to the working plans prepared for the van panchayat by the Forest Department.

d. For commercial sale or auction of forest products (fodder, grass, minor forest products, firewood, timber), the permission of the DFO must be obtained. If the value of the auctioned products exceeds Rs. 5,000, the DFO must be present. All auctions above Rs. 5,000 must be approved by the Conservator of Forests.

e. The panchayat must prepare annual budgets and submit an annual report to the DFO each year.

Table 3.1 (continued)

Subject	1931	1976
		f. Special officers appointed to supervise van panchayats must oversee at least a third of panchayats each year.
		g. Van panchayat accounts can be audited.
Rights and powers of panchayats	In general, rights and powers are similar to those of forest officials:	In general, rights and powers are similar to those of forest officials:
	1. Rule breakers are fined up to Rs. 5.	Rules 3, 4, 5, and 6 remain the same.
	2. For offenses where the fine should be higher, the panchayat can file court cases against rule breakers.	*Further Restrictions:*
	3. Fees may be levied from users for fodder, grazing, fuelwood, or construction stones.	a. All appointments by the van panchayat require the approval of the Deputy Commissioner.
	4. Grazing in the panchayat forest can be regulated, and animals that are found in the forest in contravention of rules may be impounded.	b. At least 20 percent of the area of the van panchayat is to be set aside from grazing. Land may be leased for commercial use.
	5. Cutting implements used in contravention of panchayat rules may be confiscated.	c. Fines on individual rule breakers may be compounded up to a limit of Rs. 50 with their permission and up to Rs. 500 with the permission of the Deputy Commissioner. Court cases may be filed against rule breakers.
	6. Users who break rules regularly may have their rights restricted or suspended.	d. No more than one tree may be granted to a rightholder without the written consent of more than half the panches and the stamp of Sarpanch.
	7. Guards may be appointed to monitor and enforce rules.	
Rule enforcement	All fines imposed by the panchayat are treated as government dues and recoverable using similar procedures.	Same as before.

Elections	Panchayat officials are elected for three years. New elections are to be held every three years.	Panchayat officials are elected for five years. New elections are to be held every five years.
Allocation of income	1. All income from the sale of forest products is allocated to rightholders as assigned to the van panchayat. 2. All income from the sale of resin is to be allocated in accordance with proportions determined by the Conservator of Forests (in practice it went to van panchayat). 3. Income from the sale of forest products (such as timber, resin, minor forest produce) to nonrightholders was assigned to the van panchayat.	Rule 1 remains the same. *Modifications:* a. The Forest Department is to deduct 10 percent from all gross revenues of the van panchayat as its share to meet administrative expenses. b. Net income from commercial sale and auctions is to be deposited in a Panchayat Forest Fund managed by the Deputy Commissioner. c. Twenty percent of the net income is allocated to the District Council to meet development costs. d. Forty percent of the net income is allocated to the Forest Department to maintain and develop panchayat forests. e. The remaining 40 percent of net income is allocated to panchayat to be spent on works of public utility as approved by the Deputy Commissioner.

Table 3.2
Basic statistics on the Dhauladevi and Lohaghat forest councils

Name of Forest Council	Area of Council Forest (hectares)	Distance from Road (kilometers)	Elevation (meters)	Number of Households
Pokhri, Almora	20	0.5	1,100	10
Tangnua, Almora	11	0	2,000	21
Kana, Almora	25	0	2,000	25
Kotuli, Almora	35	0.5	1,700	50
Bhagartola, Almora	63	1	1,900	70
Lada, Pithoragarh	33	1	1,750	30
Kadwal, Pithoragarh	21	0	1,700	15
Jogabasan, Pithoragarh	74	1	1,800	15
Goom, Pithoragarh	80	1	1,750	75

forests is concerned. The one exception is Jogabasan, which has only 15 households but whose forest area is 74 hectares. With nearly 5 ha of community forest per household, its residents have exceptionally high access to forest resources and possibly among the highest forest endowment in the districts of Almora and Pithoragarh. Many forest councils have an average of less than 1 ha of forest per household.

Table 3.3 is arrayed slightly differently in comparison to table 3.2. Table 3.3 starts with the council that has the smallest number of households (Pokhri). The rest of the councils are listed in order of ascending size, with the council that has the largest number of households (Goom in Pithoragarh with 75 households) at the bottom of the table. If we look at table 3.3, which provides summary figures on the operations and budgets of the nine councils, we find that the small councils are not doing as well as the large councils. This is especially true for the budgets of the councils but also to some extent for the number of times they meet each year.

Clearly, the small number of households has some major implications. Consider the first six councils. We can classify them as small since none of them have more than 30 households. The average annual number of

Table 3.3
Institutional information on the Dhauladevi and Lohaghat forest councils

Name of Forest Council	Number of Households	Year of Formation	Meetings per Year	Annual Budget for Protection (Rs)	Contribution per Household (Rs)
Pokhri, Almora	10	1989	2	200	20.00
Kadwal, Pithoragarh	15	1963	4	110	7.33
Jogabasan, Pithoragarh	15	1962	7	50	3.33
Tangnua, Almora	21	1988	4	175	8.33
Kana, Almora	25	1991	4	410	16.40
Lada, Pithoragarh	30	1970	5	350	11.67
Kotuli, Almora	50	1962	8	1,750	35.00
Bhagartola, Almora	70	1939 ˙	12	3,100	44.3
Goom, Pithoragarh	75	1962	6	1,645	21.9

Note: At the time of fieldwork, 33 Rs. equaled 1 US dollar.

meetings for the councils from Almora—Kana, Pokhri, and Tangnua—
lies between two and four. The average number of meetings for all the
six councils in the small category is just greater than four. Of the three
councils from Pithoragarh in this group, only one has a large number of
meetings—Jogabasan, with its average of seven meetings a year. The
main reason that Jogabasan has such a high number of meetings is that
it has a large forest and its members are attempting to raise funds by
selling some of the trees through the Uttar Pradesh Forest Corporation.
But because they have not received much cash yet, they have relatively
little to spend on protection. For the larger councils—Kotuli, Bhagartola,
and Goom—the average number of meetings ranges between six and 12,
with a group average of more than eight meetings a year. This average
is almost double that of the councils in the small group.

Data from the meeting records of the smaller councils indicate that
they have also been relatively lax in creating rules to guide user behavior
and ineffective in enforcing rules. Thus, while the meeting records of Bhagartola and Kotuli contain lists of rule breakers, the dates when the council forest guard detected rule violations, and the amounts levied as fines,
the minutes of meetings in Pokhri, Kadwal, Tangnua, and Kana are

bereft of these details. By looking at the records, one might conclude that no rules were ever broken in these four councils. Yet in interviews and informal discussions, the members of the councils talked about limited resources and the problems they faced in monitoring rule infractions. The absence of rule breaking in formal records is an indication of lax local supervision and enforcement (see also Agrawal, 1994, 277). Both Lada and Jogabasan in Pithoragarh, the two other small councils, are attempting to enforce some of the institutional rules for protecting the forest. They are facing difficulties, however, in raising the necessary funds for enforcement. Goom, which is the largest council in Pithoragarh, has six meetings each year. Again, its meeting records contain various details about rule violators in contrast to the smaller councils in Almora.

In part, these differences in the organizational performance among the councils may simply indicate differences stemming from age. At first glance this seems especially true of the councils in Almora. The three councils that are not doing well organizationally—Pokhri, Tangnua, and Kana—are all young. Their officials and their members may need more experience: in working with government officials, in interacting with each other, and in forming and enforcing rules. They may not yet have been able to establish a core set of procedures to guide daily activities. The data from Pithoragarh councils partially corroborate this view. At least Lada and Jogabasan, which have been in existence longer than the three small Almora councils, attempt to get together and create rules. Thus the performance of the councils may be a result of experience over time.

But overall, there are several problems with this explanation. In Pithoragarh, all the four councils were born around almost the same time (see table 3.3). If age were the primary explanatory variable, it is not clear why Goom and Lada seem to be doing somewhat better than Kadwal and Jogabasan. The Goom forest council seems to be doing far better than the other three councils in Pithoragarh despite being born at the same time. A closer look shows further problems with the explanation relying on age. Records for meetings of the Bhagartola and Kotuli forest councils are available for analysis. These records reveal that the councils met regularly and often and crafted a variety of rules right from birth. Their current organizational capacity certainly has developed over a period of time, but this cannot be taken to deploy time alone as the explanatory variable. [A more favorable institutional and political climate in the earlier period might have helped establish the authority of the older councils and may still be playing a role in their continued survival and success.] However, the current macroinstitutional environment has existed at least since 1976 when the Council Rules were modified. It is difficult to accept that the effects of a supportive environment could still be lingering. Equally important, it is also necessary to understand how the activities and the processes within the councils relate to the macro-environment rather than leaving the explanation to undefined historical changes.

A second difference that distinguishes the six small councils (Pokhri, Tangnua, and Kana in Almora, and Lada, Kadwal, and Jogabasan in Pithoragarh) from their larger counterparts is their meager budget. During the course of their existence the small councils have seldom been able to raise more than Rs. 750 a year to meet their expenses. If we examine only the protection budgets of these councils for which figures are presented in table 3.3, the situation is even worse. Whereas none of the small councils raise more than Rs. 500 a year on the average for protection, Kotuli, Bhagartola, and Goom routinely raise around Rs. 1,500 to 3,000 for safeguarding their forests. (Since all councils need money to hire a guard or must be able to raise volunteer labor from members to substitute for the guard, the level of budget and contributions from members become crucial elements in the successful functioning of the councils. Higher aggregate contributions from member households increase the capacity of the councils to hire guards and enforce rules.]

To some extent, the ability of households to contribute to the forest councils relates in a circular fashion to the condition and type of vegetation in the forest itself, making conclusive assertions hazardous. If villagers receive little benefit from the forest, they will have little incentive to contribute to protect the forest. In a vicious cycle, then, the degraded condition of forest will worsen still futher, discouraging future contributions. Too much, however, should not be made of such a connection. In a condition of generalized poverty in the hills, where few, if any, of the households can be viewed as prosperous or even reasonably well off, why do we find "institutional robustness" (Ostrom, 1990) in some cases and miss it in others?

In the case of the forest councils, the vicious-cycle explanation is somewhat off the mark. The per capita forest area in the case of all the councils is low, but no lower for the smaller councils than for the larger ones. In addition, more than a third of the residents in the hill villages, including the small villages, must initiate the process of forming the council. Most of the other villagers in our cases were willing to experiment. Villagers find significant proportions of their subsistence needs for fuelwood, fodder, and construction timber in the council forests. Thus they are quite dependent on forests for their survival. Finally, even in the smaller villages, there have been some contributions to the council coffers—all of these indicate that the problem is somewhat different from what the postulation of a "vicious cycle" suggests. It is related more to the inability of small groups of poor households to generate a surplus for protecting commonly owned and managed resources, rather than to their unwillingness.[10]

The success of the larger councils is reflected in the greater number of meetings held each year, the larger budgets and the higher levels of monitoring and enforcement, and even in a relatively higher level of vegetation in their forests. The figures in table 3.4 are revealing in this regard. The numbers for the column "Total Wood Volume (cubic meters per hectare)" show that the forests of the larger councils are in a somewhat better condition than those of the smaller councils.[11] Kotuli, Bhagartola, and Goom each have more than approximately 300 m³/ha of wood volume. The smaller councils have a lower level: on the average about 200 m³/ha.[12]

As one would expect, this is not true, however, for the number of stems per hectare. Certainly, the average number of stems for the smaller councils is about 1,160, in comparison to the average for the large council forests, over 1,650. But these averages mask variation within the group of small and large councils. Tangnua, a small council, has more than 2,000 stems per ha, but Goom, the largest council, has only about 700 stems per ha. Overall, the lack of new chir seedlings in all the Pithoragarh forests is a cause for concern. In Almora district, Bhagartola and Kotuli have a large number of stems per hectare among the large councils, but so do Kana and Tangnua.[13] Kotuli and Bhagartola also have a high volume of wood, and in their case it might be argued that the condition of

Table 3.4
Vegetation data for the investigated sites

Name of Forest Council (number of sampled plots)	Trees per Hectare	Mean Diameter at Breast Height of Trees (meters)	Mean Height of Trees (meters)	Total Wood Volume (cubic meters per hectares)	Number of Species (major species)
Pokhri, Almora (16)	1,096	0.182	9.3	265	5 (chir)
Kadwal, Pithora (16)	688	0.160	12.3	170	3 (chir)
Jogabasan, Pithora (16)	641	0.135	12.3	113	6 (chir)
Tangnua, Almora (9)	2,082	0.133	6.1	176	8 (chir)
Kana, Almora (20)	1,736	0.150	8.0	245	23 (utees, chir, aiyar, banj)
Lada, Pithora (12)	760	0.225	8.9	269	2 (chir)
Kotuli, Almora (26)	2,446	0.167	6.3	338	11 (banj, chir, deo-dar)
Bhagartola, Almora (18)	1,818	0.176	7.7	341	11 (banj, aiyar, chir)
Goom, Pithora (16)	697	0.264	9.2	351	5 (chir)

the forest is a result of better monitoring and enforcement since they have had community forests for a long time. Because the forests of both Kana and Tangnua have been under council management only for a short while, it is hazardous to venture about the large number of stems being a result of council management, especially when the records of these two councils do not provide evidence of careful management.

Implications of the Study

The salient features of the situation can be summarized. A number of villages in Kumaon compete with each other to subsist on the available forest resources. Of these villages some have formed local forest councils under the auspices of the Forest Council Rules of 1931. These forest councils have experienced varying degrees of success in protecting their forests. The per household endowment of forest resources is similar across the selected cases. But the absolute size of the councils varies, both in area and number of households. The rural context is unremittingly one of high levels of dependence on forests and low levels of income. Smaller forest councils have found relatively less success in protecting their resources. This last finding of the study is worth considering at greater length.

According to most writings that explore the relationship between collective action and group size, the probability of collective action becomes progressively bleak as group size increases. The data on nine forest councils indicate, however, that smaller groups may find it too arduous to create viable institutions that will persist over time to encourage collective action. The larger forest councils, on the other hand, found it relatively easier to create and maintain processes that would organize their members and ensure their contribution to forest protection.

Two reasons can be advanced to explain the success of larger forest councils. Each relates to protection of forests from unauthorized users and uses. To protect forests successfully in a context of generalized pressure on resources, councils need guards who will enforce rules. But guards who will monitor the condition of forests and prevent rule infringements cannot be hired without a minimum level of surplus. The smaller communities of poor peasants find it difficult to contribute even the relatively modest amounts that are necessary to hire a guard. As group size increases, it becomes easier to organize a surplus and commit it to enforcement and monitoring (Thompson, 1977; Agrawal, 1992).

Second, smaller councils also find it more difficult to prevent residents of other villages from coming and breaking rules related to forest use. In any dispute with residents of other villages, they command fewer resources that would enable persistence in imposing sanctions on rule breakers,[14] especially in the absence of adequate support from the Revenue Department and other higher authorities. If a village population cannot raise sufficient resources to hire a guard to detect and prevent rule infractions, it is unlikely to possess the resources needed either to influence higher-level government officials or to move the notoriously slow Indian judicial system to resolve disputes. Thus, on both counts—hiring a guard and influencing higher-level enforcement mechanisms—smaller councils are disadvantaged.

The finding that relatively larger groups found it easier to protect their forests successfully permits an engagement with the impressive theoretical literature on the relationship between group size and the probability of collective action. Before Mancur Olson's celebrated *The Logic of Collective Action* in 1965, Buchanan and Tullock (1962) inquired into the circumstances under which rational individuals would organize themselves to produce collective goods. Their discussion, however, assumes well defined and enforced property rights and focuses primarily on the internal dynamics of a group rather than on the results of competition between asymmetrically sized groups. In the situation in Kumaon, it is precisely the delineation of property rights over forests and their enforcement that is an issue of contention.

Olson's seminal work points to the importance of group size itself in determining whether collective action will be undertaken. According to him, "Unless the number of individuals in a group is quite small, or unless there is coercion or some other special device to make individuals act in their common interest, *rational self-interested individuals will not act to achieve their common or group interests*" (1965, 2, emphasis in original). Focusing on the internal dynamics of groups by examining the motivations of individual members, Olson shows that groups will form to supply collective goods only under restricted conditions—and that these conditions are more likely to be met in small rather than large groups. As he puts it, "The larger the group, the farther it will fall short of providing an optimal supply of a collective good" (1965, 48).

In the wake of Olson's work, a number of studies have focused on the impact of group size on collective action. Hardin (1982), for example, summarizes earlier works (Buchanan, 1968; Chamberlin, 1974; Frohlich and Oppenheimer, 1970; Guttman, 1978; Hardin, 1971) to disentangle the effects of the nature of the good, the relation between the costs of collective action and the benefits of the collective good per group member, and the likelihood of collective action. A large number of later studies have also tried to relate the possibility of collective action with group size, heterogeneity of member interests, reciprocity and interdependence, and marginal per capita returns from the provision of collective goods (Isaac, Walker, and Williams, 1994; Komorita, Parks, and Hulbert, 1992;

Massey, 1994; Oliver and Marwell, 1985, 1988; Rapoport, Bornstein, and Erev, 1989; and Yamagishi and Cook, 1993). These studies have substantially enhanced our understanding of the impact of group size on collective action and of collective action more generally.

The example of the forest councils in Kumaon, however, highlights some of the aspects of the relationship between group size and collective action that merit greater attention. The following discussion extends existing studies of collective action by making two additional points: it calls into greater focus the external dynamics of a group with other groups, and it makes a distinction between the forming of a group and achieving the objectives for which the group was formed.

Most existing studies have focused only on the internal dynamics of the group—the relationship among group members. Following Olson's forceful focus on the rational, self-interested individual as the constituent unit of all groups, later studies have also focused primarily on the individual and his or her relationship to collective action. In the process, they have ignored the impact of external relationships of a group with other groups. They have seldom considered how in a situation where different groups and their members compete for resources with other groups and the members of other groups, surely a widespread phenomenon, group size may be positively related to successful collective action, at least for some range.[15]

The logic is devastatingly simple, almost "tautological," as Hardin characterizes part of Olson's argument (1982, 38). Most villages in the Kumaon hills already exist as groups. Individuals are born into these groups. The choice they face is not whether to join a group. Rather, they must choose *not* to join a group of which they are already a member by virtue of birth. Their calculus is not about the costs of joining; rather, it is about how expensive it would be not to join. In this situation, where individuals find it costly to leave the group rather than to join, it is obvious that informal groups will exist easily. The question is why among these informal groups one would find that the larger ones are more successful.

While villagers already existed as groups before the Forest Council Rules of 1931, the passage of the Rules lowered the costs of constituting the village as a formal legal entity to protect the local forests. Government officials from the Revenue and the Forest Departments encouraged villagers to form councils. If villagers agreed to do so, they could bring forests under the control of the Revenue Department under their own control. Further, owing to the scarcity of forest products in the hills, villagers are often forced to harvest them in violation of existing rules protecting community forests. In the "drab everyday struggle" (Lenin, 1902, rpt. 1976, 93) to protect their resources from others, it is not surprising that larger councils gain greater success than smaller ones.

Larger groups are more successful in two senses. First, a group that gains in size as more villagers participate in its activities is better able to raise more resources and expend a greater monitoring and enforcement effort. Second, if there are a number of different groups, some larger than others, the larger groups are more likely to be successful.[16] Both these propositions rely on an additional distinction between organizing collective action and success in achieving the objective of collective action.

Many studies of collective action assume, almost by default, that success in organizing a group (or collective action) and success in meeting the aims for which the group (collective action) is organized are one and the same thing. Under many conditions, the distinction is unnecessary— perhaps the reason that this particular obfuscation has survived so long. Successfully organizing a march to protest abortion rights is synonymous with succeeding in the objective of organizing a march. But if the objective of the marchers is to overturn *Roe v. Wade,* then success in organizing the collective action (marching) is quite distinct from success in achieving the objective of collective action.

In the case of the forest councils, successfully forming a group to protect village forest resources is a very different proposition from succeeding in protecting the forests. Success in forming a group may come easier to smaller groups, but success in protecting resources is easier for larger groups. What we should note is that successful collective action is not just about forming groups; it is as much about being successful in achieving the objective for which the group was formed.

The above distinction is not the same as the difference between initiating and maintaining collective action. To take the example of the forest councils again, forming a council is distinct from making sure that meetings are occurring regularly, which in turn is different from protection of the local forests. The difference between initiating and maintaining collective action necessarily depends on a temporal disjunction. But the difference between organizing collective action and achieving the objective for which the action was undertaken may or may not possess a temporal dimension. Once this distinction is made, it is easy to see that while small groups may find it easier to organize themselves, a larger group may find it easier, in comparison, to succeed in its objective, especially where protection from outsiders is concerned. The logic also operates at the level of the calculating individual. Villagers, discovering that smaller groups find it harder to protect forests from rule breakers, may well calculate that it does not make sense to continue to contribute to an unsuccessful council's demands for revenue.

If it is true that as group size increases, the likelihood of successful collective action is also likely to increase (at least for some range), the natural question is whether continuing increases in size would, at some point, begin to lead to a decline in the probability of successful collective action. It seems unlikely that groups could continue to grow *indefinitely*, even if continued growth is positively related to greater success in the achievement of objectives.[17] The studied cases have little to say about the effect of extremely large size on the probability of success. But ultimately, the costs of coordination are likely to increase sufficiently that they would outweigh benefits from increases in size. The exact point at which this would take place is, however, a function of the context in which groups operate. In the Indian Himalaya, where natural factors such as uneven topography, limited water availability and arable land, and constraints on forest-products supply restrict the growth of villages beyond a certain size, the costs of coordination in existing villages are unlikely to be extremely high. Most villages are smaller than 200 households. One can then hypothesize the following: In small communities of poor users who use common-pool resources for subsistence, the likelihood of collective action to protect local resources increases as group size increases. It may however decline as the group becomes very large and creates high costs of coordination.

The latter part of the hypothesis is based on the existing literature on collective action rather than on the data from the studied cases that provide only indirect indication of what might happen to the likelihood of successful collective action as groups become extremely large. It is because coordination costs will be very large for dispersed groups that small villages are unable to join each other to form larger forest councils. For example, Pokhri, Tangnua, and Kana are more than 6 km away from each other. They lack incentives to form a large joint council.

Conclusion

In conclusion, it may be useful to point to some practical relevance of the research. The findings reported here become significant in light of the most recent trends in forest policies in a number of countries that are attempting to take recourse to community-based conservation in an effort to move away from centralized exclusionary policies that seem to have failed. In a number of statements issued between 1988 and 1995 the central Indian government and the governments of more than 17 Indian states have sought to increase local participation in the management of Indian forests (SPWD, 1992). These Joint Forest Management statements constitute a break from the colonial forest policy that had continued in most parts of India, with only a few minor changes, even after independence. Yet the changes introduced today are far more timid than the British Forest Council Rules of 1931 that this chapter examines. Most state policy statements allow local populations only a partial share in the benefits from protecting forests and do not permit them a voice in crafting the rules whereby the forests would be managed (SPWD, 1992; GOI, 1992, 1993). Without local mechanisms to ensure adequate protection, funded through local sources of revenues, it seems unlikely that the proposed cooperation between state governments and the local communities will be fruitful.

In addition, the research indicates that where groups are very small and compete for a share in local resources, their performance in protecting resources may improve if government policies create institutional incentives for smaller groups to join together. The attempts of very small groups to protect their resources may founder because of their limited ability to raise a surplus that would enable effective local monitoring and enforcement. At the same time, if small groups are highly dispersed, the external environment may prevent the formation of institutions that could help the coordination of resource management and protection.

The relevance of the research for India is evident in the context of a declining forest base and changing forest policies. The research is also significant in the context of the emerging international debate over the criticality of local communities and indigenous institutions in managing forests. The example of the forest communities in the Indian Himalaya suggests that autonomy to local communities may need supplementary arrangements that will help protect local resources by the creation of user groups that are not too small to protect their resources. Such arrangements might also encourage dispute resolution among users.

Acknowledgments

I owe this paper, in more ways than I can mention, to the immense cooperation I received from villagers in Kumaon and the Kumaon Forest Panchayat Research Team. Special thanks go to Sushree Meenakshi Shailaja and Sri Nikunj Bharati, Taradutt Pandey, and Tara Singh Bisht. Mark Baker, Clark Gibson, Margaret McKean, Elinor Ostrom, Kimberly Pfeifer, Jesse Ribot, Mary Beth Wertime, and Jonathan Lindsay provided valuable comments through the various drafts of this paper. Discussions at the Workshop in Political Theory and Policy Analysis, Bloomington, Indiana, as we worked on the IFRI instruments, proved invaluable during fieldwork. I would also like to acknowledge the cheerful help I received from Julie England and Joby Jerrells in the analysis of data for the paper. Patty Dalecki was, as usual, sterling in her editorial efforts. Grants from the Division of Sponsored Research, Tropical Conservation and Development Program, and the World Wildlife Fund made the fieldwork possible. The research was also assisted by a grant from the Joint Committee on South Asia of the Social Science Research Council and the American Council of Learned Societies with funds provided by the National Endowment for the Humanities and the Ford Foundation.

Notes

1. The causes for the emphasis on local institutions may lie in the demonstrated deficiencies of state-directed development and the inability of markets to promote sustainable use of common resources. A large literature documents the vigorous debate on the merits and problems of pursuing development and conservation

goals through state- or market-led policies. For useful introductions see Bates (1981, 1989), Wade (1990), and Wolf (1988).

2. For a discussion of the relationship between renewable resource scarcity and social tensions, see Gleick (1989), Homer-Dixon (1991), and Westing (1986).

3. See Escobar (1991, 1992), Scheper-Hughes (1992), Scott (1985), Shiva (1988), and Trainer (1985) for some critiques of market- and state-led development and conservation policies that ignore the interests of the subaltern groups. The theoretical literature that stresses the necessity of addressing local interplays of power and resistance often finds its inspiration from the works of Michel Foucault (see especially 1978, 1991a, 1991b).

4. The very largest council in the sample also finds it difficult to organize successfully, and rather than treating that as an exception, I try to adduce some reasons at the end of the chapter about why this might be the case.

5. According to Somanathan (1989), these Rules only formalized the control many hill communities had exercised over their forests before the arrival of the British. Their informal institutions were called *Lattha Panchayats*. *Lattha* means "big stick," and the name evocatively denotes the power the local community held over its members.

6. Thus they seem to meet many of the design principles that are characteristic of successful community institutions as discussed by Ostrom (1990).

7. The 30 chiefs of the councils listed a total of 97 problems. Of these, 31 (32 percent) related to the low income of their councils, 22 (23 percent) to inadequate support from higher-level government officials, and 44 (45 percent) to local-level rule infringement and problems in monitoring and enforcement.

8. The selected councils have been picked from two of the development blocks (one in Almora and the other in Pithoragarh) where I conducted research.

9. The selected sites were chosen randomly out of the 11 villages in the watershed around Jageshwar that possess their own council forests and councils.

10. Even if the problem relates to lack of incentives to contribute in the smaller communities, the larger argument in this chapter holds: smaller groups find it more difficult to organize collective action successfully.

11. Since the forest councils of Kana, Pokhri, and Tangnua have formed only recently, the condition of the vegetation in their forests, unlike the cases of Kotuli and Bhagartola in Almora and all the four councils in Pithoragarh, cannot entirely be attributed to how the council functions. But the relatively lax enforcement of rules in the six smaller councils implies that the likelihood of improvement in the condition of their forests is small.

12. The calculations for the woody biomass are extremely rough. I have taken the volume of woody biomass in a hectare as the number of trees in a hectare, multiplied by the average area of the cross-section of trees at basal height (using diameter at basal height), times the average estimated height of the trees in the hectare. Since the procedure I used is the same for all forests, the volume of woody

biomass can be taken as being comparable across the council forests rather than indicating correct absolute values.

13. All the forests in Pithoragarh have a relatively small number of trees. Part of the reason for the small number of trees in Pithoragarh is that the forests seem to contain mature chir pine (*Pinus roxburghii*). As chir matures and creates a thicker layer of needles on the forest floor, it becomes harder for other species to thrive.

14. Resources in the form of labor and monetary contributions may be necessary to either discourage local rule infractions or resolve disagreements by arbitration or civil suits.

15. Rapoport, Bornstein, and Erev (1989) do consider how differences in group size may affect the probability of collective action when such groups are competing with others. On the basis of their experimental results, they conclude that group size does not have any effect on the provision of collective action. However, the group size for their experiments varies between three and five. It seems hasty to draw the conclusion that group size does not have an impact on the probability of successful competition on the basis of such minimal variance in group size.

16. In this second sense, the proposition has also found a defense from Dahl and Tufte (1973, 20–21) in their discussion of "system capacity."

17. Of course, companies vying for a larger market share certainly believe that the larger the company's control over the market, the more successfully it will outcompete its rivals.

References

Agrawal, Arun. 1992. "Risks, Resources, and Politics: Studies in Institutions and Resource Use from India." Ph.D. diss., Duke University.

———. 1994. "Rules, Rule Making, and Rule Breaking: Examining the Fit between Rule Systems and Resource Use." In *Rules, Games, and Common-Pool Resources,* ed. Elinor Ostrom, Roy Gardner, and James Walker, 267–82. Ann Arbor: University of Michigan Press.

———. 1995. "Population Pressure = Forest Degradation: An Oversimplistic Equation?" *Unasylva* 46(2): 50–58.

Anderson, D., and R. Grove, eds. 1987. *Conservation in Africa: People, Policies and Practice.* Cambridge: Cambridge University Press.

Ballabh, Vishwa, and Katar Singh. 1988. "Van (Forest) Panchayats in Uttar Pradesh Hills: A Critical Analysis." Research Paper No. 2. Institute of Rural Management, Anand, India.

Bates, Robert. 1981. *Markets and States in Tropical Africa.* Berkeley: University of California Press.

———. 1989. *Beyond the Miracle of the Market.* Cambridge: Cambridge University Press.

Buchanan, James M. 1968. *The Demand and Supply of Public Goods.* Chicago: Rand McNally.

Buchanan, James M., and Gordon Tullock. 1962. *The Calculus of Consent.* Ann Arbor: University of Michigan Press.

Chamberlin, John. 1974. "Provision of Collective Goods as a Function of Group Size." *American Political Science Review* 68(2): 707–16.

Chambers, Robert. 1983. *Rural Development: Putting the Last First.* London: Longman.

Colburn, Forrest D., ed. 1989. *Everyday Forms of Peasant Resistance.* Armonk, NY: Sharpe.

Cultural Survival. 1993. *State of the Peoples.* Boston: Beacon.

Dahl, R., and E. Tufte. 1973. *Size and Democracy.* Stanford: Stanford University Press.

Denslow J., and C. Padoch. 1988. *People of the Tropics.* Berkeley: University of California Press.

Escobar, Arturo. 1991. "Anthropology and the Development Encounter." *American Ethnologist* 18(4): 658–82.

———. 1992. "Imagining a Post Development Era: Critical Thought, Development and Social Movements." *Social Text* 31/32: 20–56.

Foucault, Michel. 1978. *The History of Sexuality.* New York: Random House.

———. 1991a. "Governmentality." In *The Foucault Effect: Studies in Governmentality,* ed. Graham Burchell, Colin Gordon, and Peter Miller. Chicago: University of Chicago Press.

———. 1991b. "Politics and the Study of Discourse." In *The Foucault Effect: Studies in Governmentality,* ed. Graham Burchell, Colin Gordon, and Peter Miller. Chicago: University of Chicago Press.

Frohlich, Norman, and Joe Oppenheimer. 1970. "I Get By with a Little Help from My Friends." *World Politics* 23(1): 104–20.

Gleick, P. 1989. "Climate Change and International Politics: Problems Facing Developing Countries." *Ambio* 18(6): 333–39.

GOI (Andhra Pradesh). 1992. "Forest Lands—Joint Forest Management—Constitution of Van Samrakshan Samitis." GOMS No. 218, 28 August.

GOI (Himachal Pradesh). 1993. "Participatory Forest Management." No. Forest (C) 3-4/80-V, 12 May.

Guha, Ramchandra. 1990. *The Unquiet Woods: Ecological Change and Peasant Resistance in the Himalaya.* Berkeley: University of California Press.

Guha, Ranajit, and Gayatri C. Spivak, eds. 1988. *Selected Subaltern Studies.* New York: Oxford University Press.

Guttman, Joel M. 1978. "Understanding Collective Action: Matching Behavior." *American Economic Review* 68(2): 251–55.

Hardin, Russell. 1971. "Collective Action as an Agreeable N-Prisoners' Dilemma." *Behavioral Science* 16(5): 472–81.

————. 1982. *Collective Action*. Baltimore, MD: Johns Hopkins University Press.

Hecht, S., and A. Cockburn. 1990. *The Fate of the Forest*. New York: Harper Perennial.

Homer-Dixon, T. 1991. "On the Threshold: Environmental Changes as Cause of Acute Conflict." *International Security* 16(2): 76–116.

Isaac, R. Mark, James M. Walker, and Arlington W. Williams. 1994. "Group Size and the Voluntary Provision of Public Goods: Experimental Evidence Utilizing Large Groups." *Journal of Public Economics* 54(1): 1–36.

KFGC (Kumaon Forest Grievances Committee). 1922. "Report of the Forest Grievances Committee for Kumaon." Mimeo.

Komorita, S. S., C. D. Parks, and L. G. Hulbert. 1992. "Reciprocity and the Induction of Cooperation in Social Dilemmas." *Journal of Personality and Social Psychology* 62(4): 607–17.

Korten, D., ed. 1986. *Community Management: Asian Experiences and Perspectives*. West Hartford, CT: Kumarian.

Lenin, V. I. 1902. *What Is to Be Done?* (rpt. 1976). Peking: Foreign Languages Press.

Marglin, F., and S. Marglin, eds. 1990. *Dominating Knowledge: Development, Culture and Resistance*. Oxford: Clarendon Press.

Massey, Rachel Ida. 1994. "Impediments to Collective Action in a Small Community." *Politics and Society* 22(3): 421–34.

Oliver, Pamela, and Gerald Marwell. 1985. "A Theory of the Critical Mass, I: Interdependence, Group Heterogeneity, and the Production of Collective Action." *American Journal of Sociology* 91(3): 522–56.

————. 1988. "The Paradox of Group Size in Collective Action: A Theory of the Critical Mass, II." *American Sociological Review* 53: 1–8.

Olson, Mancur. 1965. *The Logic of Collective Action*. Cambridge: Harvard University Press.

Ostrom, Elinor. 1990. *Governing the Commons: The Evolution of Institutions for Collective Action*. New York: Cambridge University Press.

Ostrom, Elinor, Larry Schroeder, and Susan Wynne. 1993. *Institutional Incentives and Sustainable Development*. Boulder, CO: Westview Press.

Pant, G. B. 1922. *The Forest Problem in Kumaon*. Nainital: Gyanodaya Prakashan.

Rapoport, Amnon, Gary Bornstein, and Ido Erev. 1989. "Intergroup Competition for Public Goods: Effects of Unequal Resources and Relative Group Size." *Journal of Personality and Social Psychology* 56(5): 748–56.

Redford, Kent, and Steven Sanderson. 1992. "The Brief Barren Marriage of Biodiversity and Sustainability." *Bulletin of the Ecological Society of America* 73(1): 36–39.

Repetto, R., and M. Gillis, eds. 1988. *Public Policies and the Misuse of Forest Resources*. Cambridge: Cambridge University Press.

Richards, P. 1985. *Indigenous Agricultural Revolutions.* Boulder, CO: Westview Press.

Scheper-Hughes, N. 1992. *Death without Weeping: The Violence of Everyday Life in Brazil.* Berkeley: University of California Press.

Scott, James. 1976. *The Moral Economy of the Peasant: Rebellion and Subsistence in Southeast Asia.* New Haven: Yale University Press.

————. 1985. *Weapons of the Weak.* New Haven: Yale University Press.

Shiva, Vandana. 1988. *Staying Alive.* New Delhi: Kali for Women.

Somanathan, E. 1989. "Public Forests and Private Interests." Mimeo, Indian Statistical Institute, New Delhi.

Society for the Promotion of Wasteland Development (SPWD). 1992. *Joint Forest Management: Regulations Update, 1992.* New Delhi: SPWD.

Thomson, J. T. 1977. "Ecological Deterioration: Local Level Rule Making and Enforcement Problems in Niger." In *Desertification: Environmental Degradation in and around Arid Lands,* ed. M. H. Glantz, 57–79. Boulder, CO: Westview Press.

Trainer, T. 1985. *Abandon Affluence!* London: Zed.

Uphoff, Norman, ed. 1986. *Local Institutional Development: An Analytical Sourcebook with Cases.* West Hartford, CT: Kumarian.

Wade, Robert. 1990. *Governing the Market.* Princeton, NJ: Princeton University Press.

————. 1994. *Village Republics: Economic Conditions for Collective Action in South India.* San Francisco: ICS Press.

Westing, A., ed. 1986. *Global Resources and Environmental Conflict.* New York: Oxford.

Wolf, C. 1988. *Markets or Governments.* Cambridge: MIT Press.

Yamagishi, Toshio, and Karen S. Cook. 1993. "Generalized Exchange and Social Dilemmas." *Social Psychology Quarterly* 56(4): 235–48.

4

Successful Forest Management: The Importance of Security of Tenure and Rule Enforcement in Ugandan Forests

Abwoli Y. Banana and William Gombya-Ssembajjwe

Introduction

Uganda's forest resources are an essential foundation for the country's current and future livelihood and growth. Over nine-tenths of Uganda's energy requirement, for example, is generated by forests (Ministry of Finance and Economic Development, 1993). Forests are also important for timber and for their role in increasing agricultural productivity. They support wildlife and other forms of biodiversity vital for the country's future heritage, as well as for generating foreign exchange through a tourist industry focused on the diverse flora and fauna of Uganda.

These valuable forest resources are disappearing rapidly. The 1992 Uganda National Environmental Action Plan (NEAP) estimated that deforestation was occurring in Uganda at the rate of 500 square kilometers annually, while the United Nations Food and Agriculture Organization (FAO) (1993) estimated it to be at 650 km^2 annually.

The proximate causes of forest loss are clearing for agriculture, pitsawing, and logging for lumber, charcoal, and firewood production. However, not all forests are experiencing this problem equally; in some forests we do not find overexploitation. If we can come to understand why certain forests do not experience overuse, perhaps these lessons can help construct management schemes that are more effective and sustainable.

Among the more important independent variables that affect the level and type of consumptive utilization of forests in many settings are the security of tenure that local residents possess related to forests and the level of rule enforcement related to the use of forest resources. These variables are important because individuals who lack secure rights to forest resources are strongly tempted to use up these resources before they are lost to the harvesting efforts of others. Further, if rules regulating access and use of forest resources are not adequately enforced, the *de facto* condition becomes one of open access rather than secure tenure.

In this chapter, we argue that the condition of forests in Uganda is related to the uncertain status of land and forest tenure regimes. In our study of five forests, ranging from 60 to 4,500 hectares, we find that in those areas where a system of property rights is well known to the local population and is enforced, the condition of forests is arguably better than in those areas where locals play no part in forestry management and national laws lack enforcement (NEAP, 1992). We also find that in addition to government-enforced rules, the recognition of indigenous rights to forest-resources management leads to successful management practices.

Forest Use in Uganda

To establish the effect of the independent variables described above on the outcomes (deforestation or sustainable use of the resource), studies were conducted during the fall of 1993 in five selected sites located in Uganda's four agroecological zones (tall grasslands, short grasslands, semiarid, and highlands).

Two forests were studied in the tall grassland zone in Mpigi District about 30 km west of Kampala. Two forests from one site were included because they represented a "natural experiment" in which very similar natural forest lands were divided into two forests with different tenure regimes and use rights. One of the forests is known as Namungo Forest, which is a privately owned 40 hectare patch. Adjacent to Namungo Forest is a 1,000 ha section of the Lwamunda Forest, which is a government forest reserve. Both of these forest patches are tropical moist evergreen with closed canopies (Barbour, Burk, and Pitts, 1987) and are locally classified as medium-altitude Piptadenistrum-Albizia-Celtis, after the three typically dominant species in this area (Howard, 1991).

From the highlands agroecological zone, we studied the 1,200 ha Echuya Government Forest Reserve, located approximately 500 km southwest of Kampala in Kabale District. It is a montane forest characterized by *Arundinaria alpina* bamboo species and scattered *Dombeya-Macaranga* tree species (Banana et al., 1993a, 1993b). From the semiarid agroecological zone, we selected the Mbale Forest Reserve (1,207 ha). This forest, a savanna grassland forest characterized by *Acacia-Albizia-Combretum* tree species and *Cymbopogon afronadus* and *Hyparrhenia spp*, is located approximately 70 km north of Kampala in Luwero District (Banana et al., 1993c).

Bukaleba government forest reserve (4,500 ha), located 140 km east of Kampala in Iganga District, was selected to represent forests in the short grass agroecological zone. It is a wooded savanna grassland forest, characterized and dominated by *Combretum, Teclea,* and *Terminalia* tree species (Banana et al., 1993b).

Level of Consumptive Utilization

Local forest users consume a wide variety of forest products in all five forests. Some of these uses are legal; a great number are not. Significantly, the intensity and pattern of these consumptive uses vary across the forests.

In all five forests, local forest users are permitted to harvest forest products for subsistence use in "reasonable" quantities. Access to these forests for other benefits, such as recreation and cultural activities, is open to all local users. If forest users desire to harvest forest products for commercial purposes, however, they are required to purchase a monthly or seasonal license from the Forest Department.

The specific pattern of legal use in each forest, however, varies. In Namungo Forest, the Namungo family (the private owner) recognizes the customary rights of the local residents located at the edge of their forest for the last half century. These residents are allowed to harvest firewood, poles, craft materials, medicinal plants, water, and fruits and wild foods from the forest (Gombya-Ssembajjwe et al., 1993). To monitor the use of this forest by local residents, Namungo employs a staff. The adjacent Lwamunda Forest Reserve, which is a government forest reserve, is also used by local residents for harvesting similar products. Prior to 1981, selective logging of trees over 80 centimeters in diameter by logging companies had been permitted and carried out in both Namungo and Lwamunda Forests. Locals living near the Echuya montane forest use bamboo stems extensively for firewood, poles, thatch, and fibers. In Bukaleba and Mbale Forests, the *Acacia-Albizia-Combretum* tree species that dominate are used extensively for commercial charcoal production by the local people, and the *Cymbopogon afronadus* and *Hyparrhenia spp.* grasses are used as thatch and for grazing by local and transhumant grazers in the dry season (Banana et al., 1993a, 1993b, 1993c).

The pattern of illegal consumptive use by local people also varies widely. Table 4.1 contains data regarding illegal exploitation and disturbance collected from a random sample. In each of the forests larger than 200 ha, a random sample of 30 plots was taken from 200 ha of a forest patch that is accessible to the community and where human foraging is likely to be high. The table categorizes five types of illegal activities observed in the plots: charcoal burning, pitsawing, commercial firewood collecting, grazing of livestock, and agricultural activity.

Distinct patterns emerge from the data. The plots in Lwamunda, Mbale, and Bukaleba Forests endure considerable illegal consumption activities. Mbale, for example, bears the highest level of disturbance, with all but four out of 30 sample plots showing evidence of illegal use; the

Table 4.1
Number of sample plots with evidence of illegal consumptive disturbance ($N = 30$ per forest)

Name of Forest	Charcoal	Pitsawing	Commercial Firewood	Grazing	Farm	No Illegal Consumptive Disturbance
Namungo	1	2	2	0	0	25
Lwamunda	3	8	10	0	0	9
Mbale	10	1	5	22	4	4
Echuya	0	0	3	1	0	26
Bukaleba	0	0	12	2	5	11

Note: In some sample plots, more than one type of disturbance was observed.

grazing of livestock appears to be the most frequent of illegal activities within Mbale Forest. In the plots of Lwamunda and Bukaleba, the commercial collection of firewood seems to be the most regular illegal use, observed in at least a third of the sample plots in each forest.

Overall, about 70 percent of the sample plots in Lwamunda, Mbale, and Bukaleba forest reserves show evidence of illegal consumptive utilization of one form or another. In Namungo and Echuya Forests, however, only 20 percent of the sample plots show such illegal consumptive use in each of the five categories. In Namungo Forest, no type of illegal use appears in more than 10 percent of the plots, while in Echuya Forest, three of the five types of illegal uses were not observed at all.

To investigate how the illegal consumptive uses presented in table 4.1 affect the physical condition of the forests, physical data were collected in each of the sample plots as well. The methodology for the data collection began with the demarcation of three concentric circles in each plot. In the first circle (1-meter radius), the amount of ground cover by species was estimated. In the second circle (3-m radius), shrubs and tree seedlings were identified and their heights measured. In the third circle (10-m radius), all trees were identified, their stem diameter at breast height (DBH) measured, and their heights estimated.

It can be noted that the consumptive disturbances were not universally as high as they were observed to be in Lwamunda, Mbale, and Bukaleba Forests. Data collected for trees indicate that tree species diversity was slightly better in Lwamunda forest reserve (73 species) than in the privately owned Namungo property (64 species) (table 4.2). The higher species diversity value in the government reserve may have come about by gap formation associated with repeated selective harvesting between 1971 and 1985, when there was no effective forest management by the state because of the prevailing civil strife (Becker, Banana, and Gombya-Ssembajjwe, 1995). When large trees are harvested, they form openings in the forest where a wide variety of seedlings may become established and compete, leading to a higher species richness (Denslow, 1987).

Species diversity was generally low in all of the sites in the Savanna and Montane forest zones. The number of species observed in these zones

Table 4.2
Summary of data collected for trees in plot samples of the pilot-study forests

Forest	Area (hectares)	Species Richness	Stems per Hectare	Mean Diameter at Breast Height (centimeters)	Total Basal Area (square meters)
Namungo	60	64	362	23.4	19.0
Lwamunda	1,000	73	338	26.6	16.0
Mbale	1,207	28	164	15.0	3.0
Echuya	1,200	18	5,556[a]	4.6[a]	9.2[a]
			180[b]	20.3[b]	6.0[b]
Bukaleba	4,500	34	190	17.8	5.0

a. Bamboo.
b. Trees.

was limited to 28 in Mbale Forest, 32 in Bukaleba Forest, and 18 in Echuya Forest.

The number of stems per hectare and total basal area were significantly higher in Namungo than in Lwamunda Forest, although the distribution of different tree-size classes were not significantly different in both forests (table 4.2). Both forests were dominated by trees having a diameter range of 10 to 40 cm. Very large trees with diameters greater than 80 cm were rare, representing less than 2 percent of the trees. Tree-size class distribution was also not significantly different in Mbale and Bukaleba. Both forests were dominated by small trees having a diameter range of 10 to 20 cm. Mature trees had been harvested for firewood and charcoal. Trees were larger in Echuya Forest, where tree harvesting is prohibited.

The data demonstrate that not all forests are being used at the same rate or in the same manner by the people living near them. Degradation was not found to be as extensive in Namungo and Echuya Forests as it was in Lwamunda, Mbale, and Bukaleba Forests. These latter three forests show serious signs of open-access utilization that, if left unabated, could lead to a local fuelwood shortage, substantial forest degradation, and loss of useful biotic resources and amenities.

The Role of Tenure and Enforcement

Security of tenure of natural resources is an important issue if local communities are to use sustainably natural resources in their localities. Tenure is a set of rights that a person or some private entity holds to land or trees (Bruce, 1989). It includes questions of both ownership and access to resources. Tenure helps to determine whether local people are willing to participate in the management and protection of forests (Bromley, 1991/92).

During the colonial period, indigenous peoples' rights to harvest and dispose of trees were significantly restricted. Similarly, after independence, Uganda's forest policy, like many other developing countries, has been characterized by the strong concentration of power over forest resources in the central state apparatus, and the corresponding lack of local participation in forest and tree management.

Failure to recognize indigenous systems of forest management and indigenous rights to resources has led to

- Fewer incentives for the local communities to protect trees,
- Disincentives for local people to engage in tree planting and reforestation projects, and
- Excessive reliance by the state on punitive measures to enforce the law.

Lawry (1990) argues that where forest habitats have little economic value to local people because of restrictive access rules, sustainable local management institutions are unlikely to emerge. Incentives for conservation by local people can be improved by increasing the value of the resource to local people by, for example, granting more access rights or by granting local communities a percentage of forest concession revenues. None of these measures have been adopted by the Forest Department.

Insecurity of land and tree tenure may explain the observed general degradation of the forests throughout Uganda. A centralized state policy that is not backed with enough resources to enforce its rules has led to a condition in which most forests in Uganda are *de facto* open-access resources.

And yet insecure tenure alone does not explain the observed variance of degradation that we found in our study's forests. The most significant difference between the forests is the high level of illegal consumptive utilization of Mbale, Lwamunda, and Bukaleba Forests and the lower level of illegal use in Namungo and Echuya. To account for this variance, we turn to an explanation that features the enforcement of rules at the local level.

Although all forest reserves have clearly defined boundaries, the study reveals that monitoring is difficult and costly in Lwamunda, Mbale, and Bukaleba because these reserves are large with long borders, requiring many forest guards to monitor them effectively. The financial and human resources available to the Forest Department, however, are inadequate to carry out the task of policing these forests. In addition, the government officials (forest guards, forest rangers, and forest officers) who monitor and enforce the rules are poorly paid and, thus, not motivated to carry out their duties. As a result, forest users who choose not to comply to the rules can easily escape detection. This allows individuals to use forests illegally and, hence, leads to forest overexploitation.

The Echuya and Namungo Forests, in contrast, have a much greater level of monitoring and enforcement. Namungo Forest is small (60 ha) with short borders and a path around two sides of it. Namungo's family lives on one side of the forest and the settlements are on the other side. Since Namungo values the forest for his own rights to harvest timber (after due notification of his intention to harvest) and employs farm workers who can be forest guards for part of each day, his forest has more guards than an average government reserve. Additionally, because local residents are allowed to exercise their traditional rights to harvest forest products (such as firewood, poles, medicines, fruit, fodder, and other forest products), residents tend to protect actively the forest against outsiders who try to use Namungo Forest. Thus, the level of rule enforcement in Namungo's Forest is relatively high, both because Namungo employs private guards and also because locals enjoy strong and secure rights to products within the forest. The advantage of the forest's small size, short borders, and perimeter path around two sides helps to make monitoring more effective.

Like the more illegally used forests of this study, Echuya is a large government reserve. But certain important features of Echuya help to limit the amount of illegal consumptive use. Although subject to the same constraints on human and other resources that discourage other government guards from effectively enforcing the national rules, the Forest Department staff in Echuya has augmented its monitoring capabilities by using the help of an Abayanda pygmy community. The department allows the Abayanda the right to live within, and appropriate products from, the forest on a daily basis—rights that other local residents do not possess. Because they live within the forest, the Abayanda are in a good position to monitor who is harvesting from the forest, especially since locals are allowed by law to enter the forest only once per week (on Thursdays). Echuya's physical layout also helps protect it from overexploitation. The Kabale-Kisoro road is the only road passing through the reserve and can be patrolled easily. Thus, while Echuya is large when compared to Namungo Forest, accessibility is difficult, the level of monitoring is significant, and the likelihood of being caught harvesting illegally is quite high.

The department's reliance on the Abayanda as forest monitors is effective for three reasons. First, because the Abayanda do not live with the rest of the community, they do not fear retaliation from those they report to the Forest Department staff. Second, the Abayanda are less likely to collude with other local residents in breaking rules since there is little interaction between the two communities. Third, the Abayanda have an incentive to protect the forest on which they depend on a daily basis.

In the other three forests, actions of local people suggest that unrestricted, unplanned, and illegal exploitation—as indicated by the levels of disturbances or illegal harvest—is not effectively prevented. The officials who govern these three resources have not minimized opportunities for activities that lead toward the rapid deforestation of these sites.

To comply with the rules regulating use of a resource, local users must

• Be aware of the possible consequences of not complying with the rules;

• Understand that there is sufficient monitoring of rule compliance; and

• Observe that individuals who abstain from illegally obtaining forest products do not compete with neighbors who obtain substantial income from illegal forest products (Ostrom, 1990).

In Lwamunda, Mbale, and Bukaleba, the local people are aware that there is no effective rule enforcement. As a result, these forests are a *de jure* state property but *de facto* open access. The absence of effective management and enforcement has turned these forests into a resource that can be exploited on a first-come, first-serve basis that leads to overexploitation.

Conclusions

While it is difficult to address many of these issues with cross-sectional, rather than time-series data, this chapter has put forward a few assertions about the importance of tenure, enforcement, and forestry management at the local level in Uganda.[1] In this chapter, we argued that security of tenure and level of enforcement of rules are critical issues in forestry management. Using five cases from Uganda, we provided some evidence that supports the view that for successful forest management to be achieved in Ugandan forests, attention must be paid to both the rules that allocate property rights over forest products and the way those rules are enforced.

This chapter indicates that forest resources are more likely to be sustainably utilized if an effective structure of institutional arrangements exists that gives rise to an authority system meaningful at the local level. A government forest reserve (state property) and a private forest (private property) can be as degraded as a communal forest (common property) if there are no effective institutional arrangements and associated organizational mechanisms to monitor and enforce rules that prevent wanton harvesting of the resource (Bromley, 1991/92). Regardless of the *de jure* property regime, all forests can be *de facto* open-access regimes if there are no effective institutions and mechanisms to enforce the rules.

Insecurity of land and tree tenure discourage local participation in forest-management and forest-protection activities. This in turn increases the cost of monitoring and rule enforcement by the state. Part of these increasing costs can be met by employing locals to monitor in the place of regular national staff, as is the case in the Echuya forest reserve. But the long-term sustainability of a strategy that merely strengthens the enforcement of national laws is questionable. First, it would be difficult to replicate the situation in which a community of individuals is willing to provide monitoring services at an extremely low rate of remuneration, as are the Abayanda. Second, a great deal of tension exists between the

Abayanda population and the others living around Echuya Forest. The Abayanda, considered an inferior social group by most Ugandans, are generally treated quite poorly by the Kiga ethnic group living near the Echuya forest reserve. This social tension could vitiate the forest management scheme that uses the Abayanda as an extension of the Forest Department.

Given management institutions in which local residents have a greater stake in the resources and management of a forest, it appears that successful forestry management might endure. Namungo Forest appears to be sustainably used not only because of its guards but because community residents are allowed to use the forest according to traditional custom. This makes residents more motivated to discourage outsiders from invading the forest.

As Uganda searches for ways to manage its forests, the lessons from these five cases may be instructive to policymakers. State-centered policies appear to have failed in many Ugandan forests; the costs of maintaining a top-down institutional arrangement necessary to protect forestry resources are far too high. Alternatives that appreciate the preferences and capabilities of local communities should be weighed, not only because they appear to reduce the costs to the central state of managing numerous small forests but because they appear to be more effective in maintaining forest patches in relatively good condition.

Acknowledgments

We would like to thank the residents of the settlements around Echuya, Mbale, Bukaleba, Lwamunda, and Namungo Forests. We would also like to thank the UFRIC team members: A. Nexus, G. Nabanoga, M. Kapiriri, P. Kizito, J. Bahati, G. Mwanbu, S. Sekindi, and S. Matovu. We would also like to thank E. Ostrom and C. Gibson for their helpful comments on the manuscript.

Note

1. IFRI protocols are designed to collect data over time, so we will return to these forests in the future in our attempt to further untangle these issues.

References

Banana, Abwoli Y., Pius Kizito, Joseph Bahati, and Anne Nakawesi. 1993a. "Echuya Forest Reserve and Its Users." Uganda Forestry Resources and Institutions Center, Forestry Department, Makerere University, Kampala, Uganda.

Banana, Abwoli Y., George Mwambu, Monica Kapiriri, and Gorretie Nabanoga. 1993b. "Bukaleba Forest Reserve." Uganda Forestry Resources and Institutions Center, Forestry Department, Makerere University, Kampala, Uganda.

Banana, Abwoli Y., George Mwambu, Gorretie Nabanoga, Monica Kapiriri, David Green, and C. Dustin Becker. 1993c. "Mbale Forest Reserve and Its Users: A Site Report." Uganda Forestry Resources and Institutions Center, Forestry Department, Makerere University, Kampala, Uganda.

Barbour, M. G., J. H. Burk, and W. D. Pitts. 1987. *Terrestrial Plant Ecology.* 2d ed. Menlo Park, CA: Benjamin/Cummings.

Becker, C. Dustin, Abwoli Y. Banana, and William Gombya-Ssembajjwe. 1995. "Early Detection of Tropical Forest Degradation: An IFRI Pilot Study in Uganda." *Environmental Conservation* 22(1) (Spring): 31–38.

Bromley, Daniel W. 1991/92. "Property Rights as Authority Systems: The Role of Rules in Resource Management." *Journal of Business Administration* 20(1& 2): 453–70.

Bruce, J. W. 1989. *Rapid Appraisal of Tree and Land Tenure. Community Forestry Note 5.* Rome: FAO.

Denslow, J. S. 1987. "Tropical Rainforest Gaps and Tree Species Diversity." *Annual Review of Ecology and Systematics* 18: 431–51.

Gombya-Ssembajjwe, William, Abwoli Y. Banana, Joseph Bahati, Monica Kapiriri, Pius Kizito, George Mwambu, Gorretie Nabanoga, Anne Nakawesi, David Green, Cheryl Danley, C. Dustin Becker, and Elinor Ostrom. 1993. "Mbazzi and Namungo's Forest: A Site Report." Uganda Forestry Resources and Institutions Center, Forestry Department, Makerere University, Kampala, Uganda.

Howard, P. C. 1991. *Nature Conservation in Uganda's Tropical Forest Reserves.* Gland, Switz.: International Union for the Conservation of Nature (IUCN).

Lawry, S. W. 1990. "Tenure Policy Towards Common Property Natural Resources in Sub-Saharan Africa." *Natural Resources Journal* 30: 403-4.

Ministry of Finance and Economic Development. 1993. *Background to the Budget 1993–1994.* Republic of Uganda.

National Environmental Action Plan (NEAP). 1992. "National Environmental Management Policy Framework" (draft).

Ostrom, Elinor. 1990. *Governing the Commons: The Evolution of Institutions for Collective Action.* New York: Cambridge University Press.

United Nations Food and Agriculture Organization (FAO). 1993. *Forest Resources Assessment 1990, Tropical Countries.* Rome: FAO Forestry Paper No. 112.

5

Optimal Foraging, Institutions, and Forest Change: A Case from Nepal

Charles M. Schweik

Introduction

Over the past decade, considerable attention has been given to the subject of human-induced forest change and the depletion of specific species in forests (Myers, 1988; Aldhous, 1993; Repetto, 1988; Lovejoy, 1980; Task Force on Global Biodiversity, 1989; Norton, 1986; Reid and Miller, 1989). Often these studies take a macro view of the problem, focusing on general political or economic influences (Repetto, 1988; Richards and Tucker, 1988). Other research shifts attention to the individual and searches for deeper understanding of influential variables that drive foraging behavior. Some of these micro-scale analyses focus on the influence of institutions or rules-in-use that create or modify human incentives and behavior related to forest-product consumption (Ascher, 1995; Angelsen, 1995; McKean, 1992; Thomson, Feeny, and Oakerson, 1992; Ostrom and Wertime, 1994; Morrow and Hull, 1996).

Micro-level investigations concerned with understanding the human impacts on forest change require some capacity to quantify forest condition and some method to analyze the change. The traditional method to quantify forest condition is to (1) take a sample of vegetation using a sampling strategy using forest plot measurements, (2) calculate aggregate species abundance indicators such as density, dominance, and frequency from these data, and (3) use these indicators to describe the current status of the forested area as a whole. Plot-level analyses are also sometimes conducted (see, e.g., Umans, 1993), but usually without attention to the spatial distribution of the plots. If the researcher is *extremely* fortunate, prior data may have been collected on forest condition, and these data can be compared with newly collected measurements. General conclusions can then be made regarding the change in forest resources and the impact of current institutional arrangements and forest policies on human foraging incentives in the region.

Unfortunately, it is rare indeed to find a study location that actually has had forest condition measures taken at an earlier point in time. In most cases, especially in developing-world settings, we possess information gaps: no prior data exist on the condition of a forest we set out to study. Even in the rare circumstances where a forest inventory has been taken, the data either are not georeferenced or are georeferenced in an aggregated form. Understanding change in the resource in this context is quite difficult, for no baseline data exist for comparison.

Scholars from geography, anthropology, and other disciplines have long been aware of the informing nature of spatial relationships: yesterday's human actions often leave imprints that remain apparent in the landscape of the natural resource of today (Pickett and Candenasso, 1995; Keller et al., 1996). In instances where we lack longitudinal data we can still extract new information related to change through the study of these patterns. Unfortunately, up until very recently, our ability to capture spatial relationships has been hampered by our inability to collect accurate spatial data. The advent of differential global positioning system (DGPS) technology provides new opportunities for the accurate georeferencing of data. Armed with this new information and digital-processing capabilities supplied by geographic information systems (GIS) and spatial statistics, we can more easily collect accurate spatial data and analyze them for expected spatial patterns. A spatial analysis provides an opportunity to extract additional information about forest change in instances where no baseline forest-condition data exist.

Further, in addition to overcoming the "no-baseline-data" problem, a spatial analysis at a forest-plot level may help shed light into community dynamics—something that might be missed using data aggregated at the forest level. For example, in agrarian societies that depend on forest products for subsistence, the existing spatial distribution of an important forest-product species may reflect human foraging decision making in response to the physical geography and established harvesting institutions, rules, or social norms. Over geographic space, particular forest locations may be subject to heavier harvesting levels as foragers respond to existing community relationships and forest institutional structure. To the researcher trying to understand how community inequities and governance arrangements influence the harvesting behavior of foragers, a spatial analysis could be quite revealing.

The goal of this chapter is twofold: (1) to understand the influence of forest governance and human foraging on a particularly important tree species, *Shorea robusta,* in one empirical Nepalese context and (2) to work toward developing new institutional analysis methods—by combining recent advances in DGPS, GIS, institutional analysis, and regression—to tease out the human dimensions of forest change using cross-sectional forest-plot data.

The chapter proceeds in the following manner. First, the study site and data-collection methods are described. Second, an overview is provided on the forest governance structure at the site, and an assessment of human foraging patterns is made based on villager reports and what we witnessed in the field. Given this knowledge, three rival hypotheses are presented related to the geographic distribution of one particularly important forest species: a "no-human-influence" pattern, an "open-access and optimal-foraging" pattern, and an "optimal-foraging combined with institutional influences" pattern. Third, a traditional aggregate forest-plot analysis is presented, and it is determined that little information can be garnered to identify which hypothesized pattern is supported. A focus at the forest-plot level is required. Fourth, three plot-level multivariate regression count models are presented, one representing each hypothesis. Fifth, statistical methods are described, and results are presented. Statistical tests are conducted to determine which hypothesis is supported. Sixth, substantive and methodological implications are discussed.

The Study Site and Data Collection

In October 1994, forested areas within the Kair Khola Watershed in the Chitwan District of southern Nepal were chosen for a study of forest governance (figure 5.1). The project, a part of the International Forestry Resources and Institutions (IFRI) research program at Indiana University,

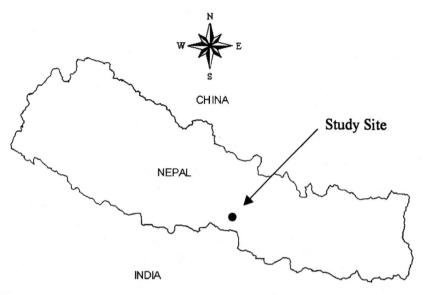

Figure 5.1
Map of the study-site location within Nepal

entailed gathering of information related to forest governance, use, and condition along with socioeconomic attributes of villagers who utilize these resources (Ostrom and Wertime, 1994). A research team comprised of Nepali researchers and the author spent six weeks in the field learning about villager foraging practices and the institutions governing forest harvesting and management. The research site falls at the juncture of the Kayar and the Shaki River systems. Figure 5.2 presents a scanned and geometrically rectified 1995 topographic map of the region. This map was created by His Majesty's Government of Nepal through interpretation of 1992 aerial photographs of the region. Grey areas designate forests; white areas reflect either degraded forest areas or areas under some agriculture regime. Four general communities exist in the study area: Milan on the west bank of where the Shakti and Kayar Rivers converge, Shaktikhor to the east along the banks of the Kayar, Latauli to north of Shaktikhor, up into the hills and Chherwan, still higher in the hills and farthest east.[1]

In general, the villagers in the west village of Milan are relatively more well off than the other communities. Many households own good land

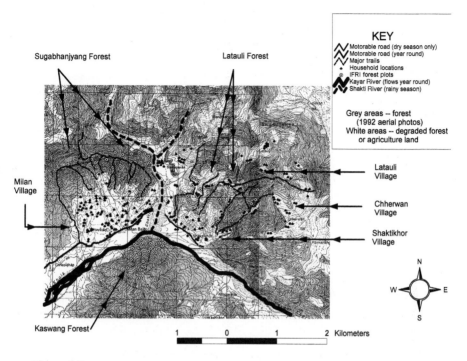

Figure 5.2
Map of the study area in East Chitwan, Nepal
Source: Survey Department, His Majesty's Government, Nepal, 1994

along the river with ample access to water resources to irrigate rice fields. Milan also exhibits more heterogeneous population when compared with the eastern communities with most members from the Chepang, Chettri, and Newar ethnic groups, but others such as Brahmin, Tamang, Gurung, and Magar are also represented. The eastern villages of Latauli, Shaktikhor, and Chherwan also are comprised of subsistence farmers living in areas where it is more difficult to irrigate given their topographic locations. Consequently, they grow other crops, such as maize, that require less water and are less commercially valuable. The scanned topographic map in figure 5.2 is also helpful, for it identifies household point locations. These point locations were also interpreted from the 1992 aerial photos and correspond reasonably well with what we witnessed in the field. In a few instances, we digitized household locations not identified in the map where we knew them to exist. These estimates are presented in table 5.1.

Table 5.1
Characteristics of the communities in the region

Community Name (from map)	Associated Village Names (from IFRI Study)	Estimated Number of Households	Forests Harvested
Milan	Sulitar, Kuwapani, Sinjali gaun, Bhandari gaun, Sewn-jaja towe, Milan Chok	110	Sugabhanjyang, Latauli, Kaswang (rarely)
Shaktikhor	Dogara	58	Latauli
Latauli	Latauli, Deurali	35	Latauli
Chherwang	Chherwang	40	Latauli

Source: Villager estimates (IFRI, 1994).

We utilized traditional plot sampling to measure forest condition. The team included one forester, one botanist, and several assistants. We utilized 10-meter radius circular forest plots for sampling. Due to the steep terrain within these forests, the team followed trails to reach 50-m altitudinal intervals. At each vertical location, a random number was used to determine the direction and the distance from the trail that the corresponding plot should be taken. Overall, 97 forest plots were sampled (figure 5.2). Data recorded include

• Soil characteristics, such as the depth of the humus layer and the depth and color of the a and b horizons;

• Tree identification, including diameter at breast height, height, and species type for each tree within the plot;

• Plot physiographic information, such as slope (in degrees, measured by a clinometer), elevation (using an altimeter), and aspect (the direction the slope faces);

• Ancillary observations, such as the existence of insect damage, signs of animal grazing, and evidence of human harvesting.

We also did something rather unusual—but soon to become more prevalent. We were fortunate enough to have two eight-channel GPS receivers and a laptop computer in the field, which allowed us to collect accurate positional data regarding these forest plots. Using differential GPS (DGPS), a technique that employs two GPS machines—one acting as a base station at a known location and the other collecting data in the field—we were able to collect forest-plot positions in longitude, latitude, and Universal Transverse Mercator (UTM) coordinate systems with an accuracy of 1 to 5 m (see Pace et al., 1995). This type of accuracy is required for a plot-level geographic analysis. These positions were converted into a GIS point coverage and are overlaid on the georeferenced map presented in figure 5.2.

Forest Governance, Use, and Hypothesized Outcomes

Villagers in each of these three communities are subsistence farmers and depend heavily on forest products for their livelihood. The term *forest*, as defined by our IFRI research program, is land area larger than .5 hectares, possessing some woody biomass, subject to the same governance structure, and utilized by at least three households. Using this definition, three forests were identified: to the west, Sugabhanjyang Forest; to the east, Latauli Forest; and to the south, Kaswang Forest (figure 5.2). Each are semideciduous *Shorea robusta* climax forests.

These forests are each designated official "government forests" and fall under the management of the district forest office (DFO). The DFO manages forests through village development committee (VDC) boundaries. The VDC is the smallest political unit in the Nepalese administration system. A VDC boundary runs directly up the Kayar River in the southwest and then follows the Shakti River northward, effectively placing the western Sugabhanjyang Forest under a different VDC jurisdiction from that governing Kaswang and Latauli.

There are three formally established DFO rules related to forest product use. First, anyone who is a member of a VDC is permitted to harvest grass, tree fodder, and deadwood from forests within that VDC to support their daily subsistence requirements. Second, live tree harvesting can be conducted only if formal permission is received from the DFO prior to harvesting. Third, a "no-encroachment" rule exists that prohibits the conversion of DFO forest to some other land use.

DFO guards stationed at VDC range posts enforce these rules. The DFO range post offices are a significant distance away from these forests: the range post associated with the monitoring of Sugabhanjyang resides approximately 14 kilometers to the southwest of Milan, and the DFO range post for Latauli and Kaswang is located approximately 16 km southeast of the village of Latauli. At each range post, approximately 10 guards are stationed. These guards are responsible for patrolling—largely on foot—a hilly, almost mountainous area that extends over 100 square km. Their task is daunting, and their effectiveness appears to be quite limited. It is not surprising that their efforts, while weak everywhere, appear to be more effective in geographic locations more easily accessible from their range post locations. Villagers report relatively few interactions with forest guards, but when they do occur they tend to be more frequent in areas along the motorable road in the western side of the study site. At one point during our fieldwork, we witnessed the DFO enforcing the no-encroachment rule. Guards destroyed the home of a villager who had encroached upon land in the western side of Sugabhanjyang near the road, and they hauled the building material away with a truck. This incident, while reportedly rare, proves that there is rule enforcement in areas reasonably close to the motorable road through Milan.

The map in figure 5.2 shows the road crossing the Shakti River and going through the eastern village of Shaktikhor. This map is deceiving, for crossing this river in a vehicle at any time of the year is quite difficult. The convergence of the two rivers at this juncture leads to a process called "the backwater effect" (Bruijnzeel and Bremmer, 1989, 64), where tremendous quantities of boulders and rocks are deposited in the Shakti riverbed. Motorable crossing is very difficult even in the dry season. The result, confirmed by villager reports, is that the monitoring of the Latauli and Kaswang Forests by DFO guards is even less frequent than in Sugabhanjyang.

Forest use by villagers in all communities is quite similar. Villagers harvest timber for construction and for tools, fodder, leaf litter and grasses for livestock and other agriculture purposes, and fuelwood for cooking and heating purposes. We witnessed extensive foraging activities during our weeks in the field. People from all villages reported that timber extraction, fuelwood gathering, and tree lopping are the major forces in what they see as a rapid depletion of their forest resources. Tree lopping is especially prevalent, which, as Metz (1990, 285) notes, significantly reduces the opportunity for species to regenerate.

While the formal DFO rules appear to be well understood, in many respects they are not followed: what we heard and witnessed in the field proved that these rules were consistently being broken. The slash and burning of forest land for agriculture perhaps is the most extreme DFO rule violation, and this practice is prevalent especially in higher locations in the hills. This aspect of the human-forest dynamics is described in much more detail in Schweik, Adhikari, and Pandit (1997). However, foraging-related violations also occur frequently. The villagers from Milan, more wealthy (relatively) and ethnically diverse, report that they harvest not only from the Sugabhanjyang Forest in their VDC, but they also lop trees for fodder and gather grasses from the eastern Latauli Forest in the neighboring VDC. This is a direct violation of the formal DFO-established rules in the region. Interestingly, the villagers from the western communities of Shaktikhor and Latauli do not seem to mind, as no complaints have been registered to the DFO range post. Even more puzzling, the villagers from the western communities of Shaktikhor, and Latauli, on the other hand, report that they forage only in the Latauli Forest. They explain that most of the year the Kayar River flows too wide for them to access the Kaswang Forest to the south and that no consideration is even given to harvesting in the Sugabhanjyang Forest in the neighboring VDC. Research team members—who lived in the villages, held numerous discussions with villagers about foraging behavior, and monitored forest-harvesting activities for nearly six weeks—confirm this behavior (Shrestha, 1996). Each side takes a "that's just the way it is" mentality when asked about these foraging patterns. It appears, then, that an unwritten social norm exists across communities that effectively permits western (Milan) villager foraging in the eastern Latauli Forest but does not allow the converse to occur. This adds additional foraging pressure on the Latauli Forest—a forest already heavily used by the Shaktikhor, Latauli, and Chherwang villagers.

A Focus on Important Product Species

If we are interested in understanding deforestation practices in a foraging community, it is most helpful to focus analysis on those species the communities find most important for their livelihood. My point here is simple but, I think, important: in any setting where foraging levels are high, the severity of the deforestation will manifest itself *first* in the distribution (or lack thereof) of the particular species that contributes most to villagers' daily subsistence requirements. When asked, the people of all three villages mentioned that *Shorea robusta* was by far the most critical tree species for supplying timber, fodder, and fuelwood needs.

Hypotheses Related to Patterns in the Distribution of *Shorea robusta*

We now can develop hypotheses related to *Shorea robusta* patterns given what we know about the forest governance structure and monitoring capacity, forest-product use, geographic locations of households, and community relationships. Three rival hypotheses exist.

Hypothesis 1: There is little or no evidence of human overconsumption of *Shorea robusta*. The forest is regenerating at a rate greater than, or equal to, what is extracted.   The first possible pattern is one of a "sustained" forest ecosystem where forests are able to regenerate at a rate faster, or equal to, the rate of what humans are removing. The pattern of *Shorea robusta* in any of the forests would be no different than what would be found in a comparable forest in a similar ecological setting that has not been subjected to human harvesting—what is sometimes referred to as a *reference forest*. Each particular species follows its own "naturally induced" distribution over the topography. Figure 5.3 describes this landscape. The likelihood that this type of pattern exists in Sugabhanjyang or Latauli is doubtful, however, given that villagers from all communities—the very people who know these forests best—report that these two forests have been significantly degraded over the past 20 years.

Hypothesis 2: *Shorea robusta* are being removed at a rate faster than the forest can regenerate. The pattern of depletion will reflect an open-access situation and a process of optimal foraging.   The second possible pattern that might exist is one that reflects an open-access situation where human decision making and harvesting are driven simply by optimal-foraging strategies. Optimal-foraging theory depicts human foragers as actors who maximize their net rate of return of energy per unit of foraging time (Smith, 1983). While a number of alternative theories on foraging decision making exist (Smith, 1983, 627), they all characterize the forager as a person who strives to minimize his or her search time and effort (Hayden, 1981; Winterhalder, 1993). If humans harvest important product

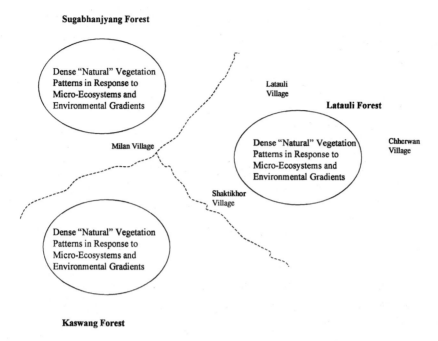

Sugabhanjyang Forest

Dense "Natural" Vegetation Patterns in Response to Micro-Ecosystems and Environmental Gradients

Latauli Village

Latauli Forest

Milan Village

Dense "Natural" Vegetation Patterns in Response to Micro-Ecosystems and Environmental Gradients

Chherwan Village

Shaktikhor Village

Dense "Natural" Vegetation Patterns in Response to Micro-Ecosystems and Environmental Gradients

Kaswang Forest

Figure 5.3
Expected patterns in a "sustained-forest" scenario

species at a faster rate than it regenerates naturally, optimal foraging predicts that lower numbers of these species will be found in locations easily accessed by humans (such as a short distance away from the village, near a path, or at a low elevation). In this setting, then, we would expect the number of *Shorea robusta* trees to be higher in number in those areas further away from villages and at higher altitudes where it is more difficult to traverse. Consequently, we would expect Kaswang to exhibit species distributions reflecting no or very low foraging activities, given that it is well protected from foraging by the river systems and few human settlements exist within or near it. We would also expect the northern part of Sugabhanjyang to be relatively untouched by humans, given that it is high up in the hills and no villages exist to the north. Alternatively, the southern half of the Sugabhanjyang Forest near Milan should be relatively hard hit in terms of foraging pressure, as well as in the eastern side of the Latauli Forest, where it is completely surrounded by settlements. But optimal foraging would predict that the forest hardest hit in terms

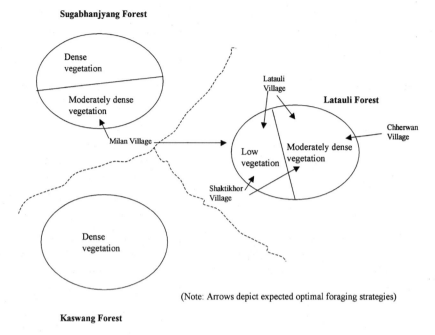

Figure 5.4
Expected patterns as a result of open access and optimal foraging

of *Shorea robusta* extraction would be the western side of the Latauli Forest—the part in the center of the map that sits in between the three villages of Milan, Shaktikhor, and Latauli. A graphic depiction of the expected foraging patterns in an open-access, optimal-foraging situation is provided in figure 5.4.

Hypothesis 3: *Shorea robusta* are being removed at a rate faster than the forest can regenerate. The pattern of depletion will reflect a process of optimal foraging altered by the geographic configuration of effectively enforced institutions. Smith (1983) reports that empirical studies testing optimal-foraging theory have revealed some instances where human foragers are selective in their utilization of available resources. Other studies have revealed foragers who exhibit much less concern. Smith also states that there is little agreement in the anthropological community over these foraging differences (Smith, 1983, 628–29). While not stated specifically, Smith's discussion alludes to the importance of community relationships and the important role institutional arrangements play in the influence of human foraging patterns and their efforts for natural-resource preservation.[2]

Ostrom (1990) extends Smith's argument by emphasizing the role institutional arrangements play in altering the incentives humans face in their decision-making context. Institutions in this context refer to the property rights and rules-in-use that govern the harvesting of a particular species or particular areas (what we might refer to as *management units*) within a forest. The forests in this particular case are, to a significant degree, open access, leading to the expectation that optimal foraging patterns will be prevalent. But the possibility exists that foraging decision making over the years may have been altered by what I refer to as the past and present "institutional landscape" configurations (Schweik, 1997). These institutional landscapes, albeit weak (see Schweik, Adhikari, and Pandit, 1997, for more detail), still may have altered human foraging behavior to some limited degree. In such an instance, the pattern in the landscape would reflect a new optimal foraging calculus, where the decision to harvest or not to harvest at a particular location includes consideration of rules, rule penalties, and the likelihood of getting caught.

In the site description presented earlier, there exist two primary institutions that appear to be somewhat influential in driving human decision making away from what optimal foraging might predict: the monitoring practices along the road in Milan and the established social norms that exist between the Milan and Shaktikhor communities. In this case, DFO guard monitoring appears to be relatively ineffective, with the possible exception of forested areas adjacent to the road through Milan. This more-prevalent forest monitoring of DFO guards in the west could add more incentive for the Milan villagers to harvest up into higher regions of Sugabhanjyang and across the river in Latauli *in locations that are not visible from the road.* This, in conjunction with the interesting social dynamic we discovered—the unwritten or accepted rule that allows villagers of Milan to harvest in the Latauli Forest but not vice versa—places added pressure on the eastern side of the Latauli Forest. Thus, in a setting where both optimal foraging and these institutional-induced incentives are present, we would expect a landscape produced that reflects more of

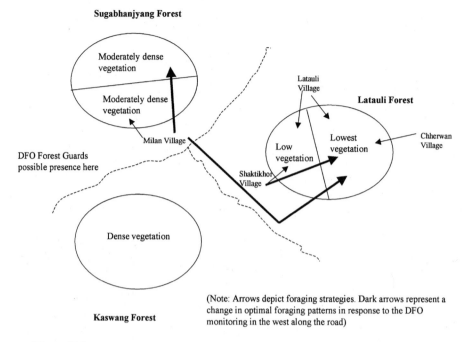

Sugabhanjyang Forest

Moderately dense vegetation

Moderately dense vegetation

Milan Village

DFO Forest Guards possible presence here

Latauli Village

Latauli Forest

Chherwan Village

Lowest vegetation

Low vegetation

Shaktikhor Village

Dense vegetation

(Note: Arrows depict foraging strategies. Dark arrows represent a change in optimal foraging patterns in response to the DFO monitoring in the west along the road)

Kaswang Forest

Figure 5.5
Patterns produced as a result of optimal foraging combined with geographic institutional influences

a continuous degradation of the forest—a depletion trend—as one moves from the west to east (figure 5.5).

Hypothesis Testing with Traditional Forest-Condition Measures

Can the traditional "aggregated" analyses of forest-plot conditions provide support for one of the three hypotheses indicated above? Figures 5.6 to 5.11 provide aggregate plot analyses for each of the three forests. Figures 5.6 and 5.7 present a comparison of the mean diameter at breast height (DBH) and the mean height of *Shorea robusta* and four other species deemed highly valuable by the villagers in the region: *Nycthanthes arbortristis* (Parijat), *Adina cordifolia* (Karma), *Lagerstroemia parviflora* (Botdhainyero), and *Terminalia tomoentosa* (Saj). While there is some fluctuation in mean DBH between forests for particular species, nothing strikingly different is identified in this comparison.

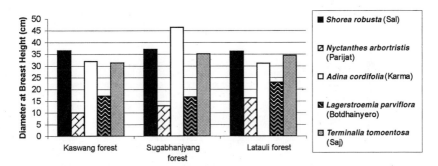

Figure 5.6
Mean diameter at breast height for preferred forest-product species

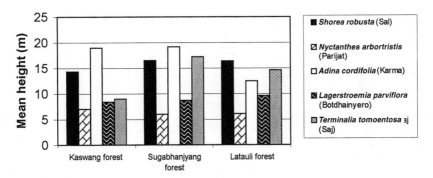

Figure 5.7
Mean height for preferred forest-product species

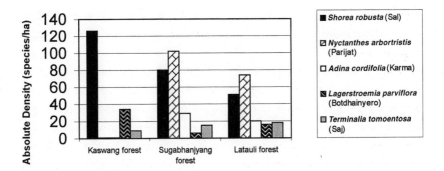

Figure 5.8
Absolute density (species per hectare) of important product species

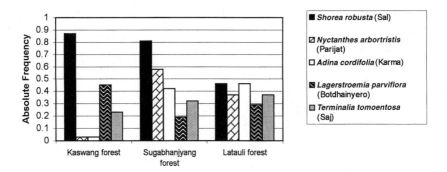

Figure 5.9
Absolute frequency of important tree species

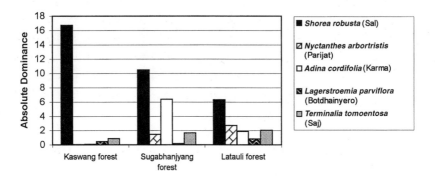

Figure 5.10
Dominance of important tree species

Figures 5.8, 5.9, 5.10, and 5.11 provide a comparison of absolute density, frequency, dominance, and species importance values of these species across the three forests.[3] Across all indicators but mean height, Kaswang reflects a much higher presence of *Shorea robusta* compared to the Sugabhanjyang and Latauli Forests. This supports the contention that Kaswang is subject to significantly less foraging as villagers suggested in the field. We would expect a *Shorea robusta* climax forest, left relatively untouched by humans, to exhibit high values in these indicators for the *Shorea robusta* species. But while it is clear that Sugabhanjyang and Latauli are comprised of differing levels of vegetation than Kaswang, this is about the only definitive conclusion we can make. We cannot easily identify

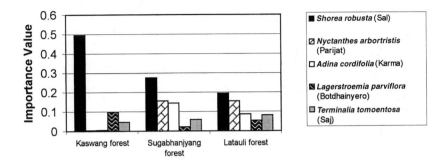

Figure 5.11
Important values for preferred tree species in the forests of Shaktikhor study area

from this one-time-point aggregate data whether Latauli or Sugabanjyang follow patterns of optimal foraging in figure 5.4 or optimal foraging coupled with institutional influences in figure 5.5. The lower measures found in Latauli and Sugabhanjyang could be a result of purely biophysical differences such as topography. In other words, Latauli could have *always* exhibited fewer *Shorea robusta* individuals than its neighboring forests.

Further Testing of the Hypotheses: Three Forest-Plot Event-Count Models

The argument made earlier is that by giving extra consideration to spatial relationships and testing for these factors we can improve our understanding of forest change in instances when baseline forest-condition data are unavailable. Three nested models will be used to test the hypotheses articulated above using multivariate regression. Model 1 is associated with Hypothesis 1. It contains abiotic and biotic factors considered important for *Shorea robusta* growth where human foraging disturbance is minimal (figure 5.3). Model 2 tests Hypothesis 2: patterns of human optimal foraging will be found in the distribution of *Shorea robusta* over the landscape (figure 5.4). Model 2 requires the control for abiotic and biotic variables of Model 1 but has additional parameters included to capture the influence of human optimal foraging. Similarly, Model 3 tests Hypothesis 3: the geography of rules and enforcement of these rules (figure 5.5) shift optimal foraging patterns to the eastern side of figure 5.2. Model 3 requires the control of all variables from Models 1 and 2 plus a parameter capturing the institutional pressures. The three models, variable operationalization, and expected signs are summarized in table 5.2.

The Dependent Variable: A Measure of *Shorea robusta* Abundance

Spatial analysis requires forest plots to be the unit of analysis rather than aggregate forest measures. We argued earlier that a focus on important forest-product species is a useful endeavor for identifying deforestation in cross-sectional data sets. Given the extreme importance of *Shorea robusta* to the villagers in these communities, the dependent variable for each of the models is a count of the number of this type of tree in a plot. A count provides a simple but useful measure of species abundance.

The Independent Variables: Factors That Influence Where *Shorea robusta* Exists

Model 1 Variables: Abiotic and Biotic Factors   Each forest plot contains physiographic characteristics that influence the capacity for particular species to grow in its environment. These abiotic characteristics include plot steepness, aspect, elevation, and soil type and condition and can play a tremendous role in the amount and type of vegetation that grows in a particular plot (Spurr and Barnes, 1992; Schreier et al., 1994).

Plot steepness is operationalized as a continuous variable measured in degrees. Figure 5.2 reveals the hilly topography of the site. In areas of extreme topographic variation, we would expect that as slope in degrees goes up, the number of *Shorea robusta* trees in a plot will go down.

Aspect captures the direction a plot faces. In one study of forests in the middle mountain region of Nepal, Schreier and colleagues (1994, 148) assume high-elevation north-facing slopes to be moist and cool, and low-elevation south-facing slopes to be hotter and dryer. But the subtropical forests studied here reside in the southern Siwalik hill region of Nepal, and slope-sunlight differences may be less pronounced. It is not clear what role, if any, aspect plays in determining where *Shorea robusta* exist. It is

Table 5.2
Three imbedded models, variable operationalization, and expected relationships

	Variable	Operationalization	Expected Sign
Dependent variable	Number of *Shorea robusta* trees in plot	Count of *Shorea robusta* trees (DBH ≧10 cm) in the plot	N/A
Hypothesis 1 variables (abiotic and biotic factors)	Steepness of plot	Steepness in degrees	(−) As steepness goes up, the number of *Shorea robusta* trees goes down.
	Aspect or orientation of plot	Degree of southness (0–4). A 0 represents a north-facing slope; a 1 represents either a northwest- or northeast-facing slope; a 2 represents an east- or west-facing slope; a 3 represents a southwest- or southeast-facing slope; a 4 represents a completely south-facing slope.	This is an unknown but potentially important relationship and should be included in the model.
	Depth of A and B soil horizons	Measured in centimeters	This is an unknown but potentially important relationship based on work by Burton, Shah, and Schreier (1989).
Hypothesis 2 variables (human optimal foraging)	Number of households within a 1-km radius of plot	Count	(−) As the number of households increases, the number of *Shorea robusta* trees should decrease.
	Average distance to plot of 1-km households	Average distance (in meters) of households within a 1-km radius of plot	(+) As the average distance increases, the number of *Shorea robusta* trees in a plot should also increase.
	Plot elevation	Measured in meters	(+) As elevation increases, *Shorea robusta* also increases.
Hypothesis 3 variables (institutional pressures)	x UTM coordinate	Measured in meters; UTM easting coordinate; subtracted the average to establish a 0,0 coordinate in the middle of the map	(−) As one moves east, the number of *Shorea robusta* trees is expected to decrease.

included in Model 1 to determine whether it has an effect and is operationalized as a categorical variable following the assumptions of Schreier et al. (1994). A 0 represents a north-facing slope; a 1 represents either a northwest- or northeast-facing slope; a 2 represents an east- or west-facing slope; a 3 represents a southwest- or southeast-facing slope, and a 4 represents a completely south-facing slope.

Elevation is another parameter thought to influence where particular species grow (Spurr and Barnes, 1992). *Shorea robusta* is known to reside at elevations as high as 1,250 m (Storrs and Storrs, 1990). The highest plot taken in our sample is 830 m. Therefore, theoretically, elevation is not needed in the model to capture altitudinal effects on *Shorea robusta* because there should be none.

Soil nutrients, moisture, and physical composition also affect the character and growth of vegetation (Spurr and Barnes, 1992; Burton, Shah, and Schreier, 1989). Four soil horizons are typically analyzed: O (humus or ground litter layer), A (between 0 and 20 cm), B (20 to 50 cm), and C (> 50 cm). Burton, Shah, and Schreier (1989, 398) studied soil conditions in degraded and undisturbed forests only a few kilometers south of this study site and report variation in A horizon samples but few differences in C horizon samples. During our fieldwork, we collected A and B horizon depth in each plot. This is included in Model 1 to capture its potential influence on *Shorea robusta* growth. Just what relationship to expect between these soil depths and existence of *Shorea robusta* is unknown. Other estimates of soil condition, such as soil color and texture, were also collected but exhibit little variation across plots.

Several biotic parameters were considered but ultimately not included in Model 1. The proximity of neighboring *Shorea robusta* seed trees often determines whether a tree will grow in a particular plot. The seed of a *Shorea robusta* tree is winged (Storrs and Storrs, 1990), and these seeds can travel great distances by wind. Therefore, given that these are all *Shorea robusta* climax forests (see figures 5.6 to 5.11), it is assumed that each forest plot has an equal likelihood of having *Shorea robusta* seed trees somewhere in its vicinity and that this factor need not be specified in Model 1.

Animal foraging may affect the fate of many seedlings. In contrast, species of no interest to animals may continue to survive or even thrive.

While animal foraging is probably an important parameter, the grazing of livestock is closely related to the location of households, and therefore its influence will be captured through optimal foraging variables specified later. The influence of other animals is assumed to be a random event.

Competition from other species is another potential parameter for Model 1. A measure of competing biomass was included in earlier runs of this model. However, after further consideration it was determined that this variable follows more of a simultaneous relationship with the count of *Shorea robusta* than a causal factor. For example, the abundance of rival species may be the result of opportunities provided because *Shorea robusta* was harvested. For this reason, competing biomass is not included in Model 1.

Model 2 Variables: Human Optimal-Foraging Pressures   Now we turn to the challenge of how to best operationalize independent variables that capture optimal-foraging pressure on a plot. Significance of such variables would lend support to Hypothesis 2 (the pattern described in figure 5.4). Five operationalization options exist.

The ideal method to capture plot foraging efforts would be to measure the distance and steepness of the trails leading to each plot from village centers or individual household locations. This approach proved to be impractical because the trails on the current map do not accurately reflect what we witnessed in the field. The topographic map in figure 5.2 reveals only a few trails; yet from field experience, we know there exists an elaborate series of trails within each forest. This operationalization approach was rejected.

A second plausible method would be to calculate a straight-line distance from each plot to the center of a village or villages. Using GIS functionality, this is a relatively straightforward task, *if* one can define village centers. This is extremely difficult in this case, for households are scattered throughout the landscape (figure 5.2). This approach was also rejected.

The third and fourth methods to quantify optimal foraging pressure involve developing a count of the number of households within a certain distance from each plot and developing an aggregate measure of how far these households are from each plot. A relatively accurate depiction of household locations existed on the 1995 topographic map interpreted from recent aerial photographs. These household point locations were digitized, and some additional household points were added from our field experience. While in some instances foragers may travel beyond 1-km distance, the assumption is reasonable given what we witnessed in the field and the hilly terrain. The Arc-Info GIS "pointdistance" function calculated the distance between each forest plot and each household point falling within a 1-km search radius. An average distance for all houses within the 1-km circle of the plot could then be calculated. Plots with a larger number of households in the 1-km circle are expected to have low counts of *Shorea robusta* trees. We would also expect that the shorter the household average distance, the more foraging pressure the plot is subjected to and the fewer *Shorea robusta* trees will be found.

The fifth and perhaps easiest method to capture a component of optimal foraging pressure relates back to the discussion of plot elevation. It was stated earlier that there is no theoretical justification to include elevation in the model for purely biophysical reasons. But elevation could be important from a human-foraging perspective because it captures the altitudinal harvesting effort required to get to and from a particular plot from household locations. Figure 5.2 shows that the majority of the villagers in this region live in the lowlands near riverbeds. Optimal foraging would predict that the villagers would tend to avoid making a trek from the riverbed to high-altitude locations for tree products if at all possible. From this discussion, the viable optimal foraging parameters for Model 2 are (1) the number of households within a 1-km radius capturing household pressure, (2) the average distance of households within the 1-km radius to the plot, capturing a distance component of foraging effort, and (3) plot elevation, capturing the altitudinal component of foraging effort.

Model 3 Variables: Institutional Influences   The question now turns to how to develop a variable that captures the effect of reasonably well-enforced rules in the western side of figure 5.2 and the more limited enforcement of rules in the eastern side on optimal foraging patterns to operationalize Hypothesis 3. One method for capturing this influence would be to assign a categorical variable to each plot that is our own assessment in the field of the likelihood (e.g., high, medium, or low) that this plot is well monitored by DFO guards. This type of operationalization, however, is subjective and would lose precious degrees of freedom in statistical implementation. For these reasons, this method was rejected.

However, given that the predicted pattern is one where *Shorea robusta* depletes as one moves farther east, another technique is a possibility. Geographers have applied coordinate systems as independent variables—what is commonly referred to as *trend surface models*—to capture trends across landscapes. We can therefore utilize the UTM plot coordinates collected by DGPS. Average x and y coordinates were calculated and subtracted from each point to establish 0,0 origin—near the point where the Shakti and the Kayar Rivers converge. The x and y coordinates represent the distance, in meters, in an east-west and north-south direction from this origin. Hypothesis 3 anticipates a trend moving from west to east (figure 5.5), so only the x UTM coordinate of a plot is required. If the hypothesized eastern trend exists, x UTM is expected to exhibit a negative sign signifying depletion in the eastern direction. Model 3 includes this variable and controls for all other variables specified for Models 1 and 2.

Statistical Methods and Results

The nested models representing the rival hypotheses will be estimated and compared using multiple regression. The dependent variable, number of *Shorea robusta* in a plot, takes on values from zero to some positive integer. Traditional ordinary least squares (OLS) regression could be applied to estimate the influence of the independent variables on this event count, but it has been shown that such an approach yields inefficient, inconsistent, and biased estimates (King, 1988; Long, 1997). Further, the OLS assumption of normally distributed residuals is incorrect when counts of biological phenomena are being estimated (Ludwig and Reynolds, 1988). Scatterplots were made of each independent variable versus the dependent variable, and the results confirm nonlinear relationships. Several nonlinear-count data models (e.g., Poisson, negative binomial) are potentially more appropriate, and the choice depends on the distributional assumption made regarding the presence or absence of *Shorea robusta*.

Ludwig and Reynolds (1988) report that counts of species usually follow one of three types of spatial arrangements: random, clustered, or uniform. In the case of a random dispersal of species, each plot has an equal chance of hosting a *Shorea robusta* individual. In such random patterns, the variance will be very close to the mean in value, and the Poisson distribution is appropriate (Ruser, 1991; Long, 1997). The second pattern, a clustered pattern, is commonly found in biological studies. Clustering will result, and a large number of plots where no *Shorea robusta* individuals exist will be identified. The variance in a clustering pattern will be greater than the mean. In these instances of overdispersion, the negative binomial distributional assumption is more appropriate. The third pattern often identified is a uniform pattern, where almost every plot exhibits the same number of *Shorea robusta* individuals. In these spatial patterns, the variance will be less than the mean (Ludwig and Reynolds, 1988).

A reference forest is required to identify the "natural" distribution of the *Shorea robusta*. A reference forest is a forest that (1) adequately represents the other forests of interest and (2) is generally undisturbed by human activity. Kaswang satisfies the above two conditions. The Nepali foresters identified all three forests as *Shorea robusta* climax forests. Moreover, Kaswang's natural protection by the river systems suggests that it is the least impacted by human activities. The aggregate measures in figures 5.6 to 5.11 confirm these suspicions. For these reasons, the 31 forest plots sampled in the Kaswang Forest are treated *separately* as the reference forest to determine a natural distribution of *Shorea robusta*. The multiple-regression models then utilize the other 66 forest-plot data from only the Sugabhanjyang and Latauli Forests.

A variance-to-mean ratio or index-of-dispersion test (Ludwig and Reynolds, 1988) helps determine the appropriate distributional assumption of the *Shorea robusta* count for the Kaswang Forest (table 5.3). The value for the chi-squared statistic (df 30) is larger than the critical value at the .01 probability level, implying that *Shorea robusta* in natural settings follows a clumped pattern (variance is greater than the mean). In such cases of overdispersion, the negative binomial distribution is appropriate (King, 1989a, 1989b; Long, 1997). Three negative binomial regression models, each representing the three rival hypotheses, will be compared.[4]

Table 5.3
Chi-square test of the index of dispersion of *Shorea robusta* trees in the Kaswang forest

Average number of individuals per plot	5.968
Number of plots	31
Variance	12.644
Index of dispersion (variance/mean ratio)	2.119
χ^2 statistic $[\chi^2 = \text{ID}(N - 1)]$	63.562[a]

a. Significant at the 99 percent level of confidence.

Several variables (such as aspect, A and B horizon depth, and specific operationalization of optimal foraging variables) are of unclear theoretical importance. Consequently, five models, not three, are actually estimated. The approach follows ideas posed by Leamer (1983, 1985) where model construction includes focus and doubtful variables. *Focus variables* are those known to be theoretically important or of particular interest to the researcher. *Doubtful variables* are those of possible importance but unclear based on prior work. All model results are provided in table 5.4 so that readers can make their own comparative judgments. Discussions of Models 1A, 1B, 2A, and 2B will concentrate primarily on whether inclusion of doubtful variables captures theoretical concepts correctly and should remain in the models. Interpretations of coefficients and model comparisons will be left for the discussion of the full model, Model 3.

Models 1A and 1B represent two alternative specifications for the abiotic and biotic factors argued in Hypothesis 1. In Model 1A, plot steepness is found to have a negative influence on the existence of *Shorea robusta* species. The expected relationship holds, and it is not surprising that this parameter is statistically significant. It is difficult for larger trees to root effectively in steep terrain. One of the doubtful variables, plot aspect, is found to be not statistically significant. This is not surprising. We witnessed *Shorea robusta* on all types of terrain facing all types of directions. Similarly, no statistically significant relationship appears to exist between another doubtful variable, depth of the A and B horizon, and the number of *Shorea robusta* trees in a plot. *Shorea robusta*, being the climax species of these forests, may be robust in its ability to grow

Table 5.4
Negative binomial coefficients for three foraging models (dependent variable is number of *Shorea robusta* trees in forest plots)

Independent Variables	Model 1A (Pattern 1)		Model 1B (Pattern 1)		Model 2A (Pattern 2)		Model 2B (Pattern 2)		Model 3 (Pattern 3)	
	Coefficient	IRR	Coefficient	IRR	Coefficient	IRR	Coefficient	IRR	Coefficient	IRR
Abiotic and biotic factors Plot steepness	-.0382[b] (.0160)	.9626	-.0369[b] (.0167)	.9637	-.0374[b] (.0154)	.9633	-.0336[a] (.0152)	.9670	-.0395[a] (.0142)	.9613
Plot aspect	-.1084 (.1462)	.8972								
A & B horizon depth	-.0312 (.0201)	.0196								
Human optimal foraging Number of households within 1 km of plot					.0121 (.0088)	1.012				
Average distance of 1-km households to plot					.0029[c] (.0016)	1.002	.0019 (.0015)	1.002	-.0012 (.0016)	.9988
Elevation					.0047[a] (.0018)	1.004	.0032[b] (.0015)	1.003	.0038[a] (.0014)	1.004
Institutional influences x UTM coordinate									-.0005[a] (.0002)	.9995
Intercept	3.079 (.8769)		2.1112 (.6755)		-3.184 (2.027)	*	-1.2628 (1.478)		.8618 (1.613)	
Log-likelihood	-121.2471		-122.8021		-118.4093		-119.3736		-116.1195	

Note: Numbers in parentheses are standard errors.
a. Significant at the 99 percent level of confidence.
b. Significant at the 95 percent level of confidence.
c. Significant at the 90 percent level of confidence.

in a variety of soils. From this analysis the doubtful variables, aspect and A and B horizon depth, do not appear to be important factors related to *Shorea robusta* growth. Consequently, a more parsimonious Model 1B is estimated to test Hypothesis 1.[5]

Model 2A adds the three optimal foraging parameters and represents Hypothesis 2. In this model, elevation is statistically significant at the 99 percent level of confidence and exhibits the expected sign. Given that *Shorea robusta* is known to grow in elevations much higher than any plot in this study, the elevation parameter undoubtedly captures an effort component of optimal foraging. Average distance of 1-km households to the plot has the expected sign and is significant at the 90 percent level of confidence. This parameter measures the trekking distance effort required to move from a household to the plot and back. The third optimal foraging variable, the number of households within 1 km of plot, is not statistically significant and exhibits the opposite sign than is theoretically expected. This variable was intended to capture the amount of household pressure on a particular plot but is probably an inadequate measure. Since it is calculated as a straight-line distance, it most likely does not adequately take into account the intricate trail network that exists. For this reason, this third variable is removed, and Model 2B is estimated. This revised model represents the parameters required for Hypothesis 2 (figure 5.4) and will be revisited in later statistical tests between models.

Model 3 adds institutional influences (Hypothesis 3, figure 5.5). Note that the other two models (1B and 2B) are imbedded and the x UTM coordinate is added to capture the anticipated influence of a foraging shift to the eastern side of figure 5.2. Plot steepness, elevation, and the x UTM coordinate are all statistically significant at the 99 percent confidence level in Model 3 and exhibit theoretically expected signs. One optimal foraging parameter, average household distance within 1 km, becomes statistically insignificant in this model.

With Model 3 representing the full nested model, we can interpret coefficients. Caution is required given that the results from negative binomial regression cannot be interpreted in the same manner as they would be if they were produced by an OLS regression. One of the most intuitive ways of interpreting these results is by creating the incident rate ratio (IRR).

IRRs can be easily interpreted as a percentage of growth or decline in the dependent variable due to a one-unit change in the independent variable, controlling for everything else. Model 3 coefficients can be interpreted as follows: holding all else constant, a one-degree increase in steepness will result in a 3.87 (100*[1-IRR] or 100*[1-.9613]) percent decrease in the expected number of *Shorea robusta* trees. Similarly, every 1-m increase in elevation increases the expected number of *Shorea robusta* trees by .4 percent (100*[1-1.004]). Lastly, the negative coefficient for the x UTM parameter suggests that for every 1-m shift east, the expected number of *Shorea robusta* trees will decrease by .05 percent (100*[1-.9995]). In other words, holding abiotic, biotic, and other optimal foraging parameters constant, as one moves east on the map, the number of *Shorea robusta* trees found in plots decreases. The percentage decrease is relatively small because 1-m on the map is a fine unit.

To verify that no multicollinearity problems exist, pairwise correlation coefficients were estimated for all parameters in Model 3 (table 5.5). Only the x UTM and the average household distance variables show any potential signs of being related, and this relationship is not very strong. Given that they both are calculated from the GIS grid, it could be that a small component of the x UTM coordinate effect is picking up a portion of the effects of the traditional optimal foraging process. However, that alone does not explain why the x UTM coordinate is so strongly related—much more so than the average household 1-km distance variable—to the count of *Shorea robusta* trees.

Table 5.5
Pairwise correlation coefficients for Model 3 parameters

	Steepness	Average Distance of 1-km Households	Elevation	x UTM Coordinate
Steepness	1.0000			
Average distance of 1-km households	−0.1840	1.000		
Elevation	0.0184	0.0880	1.000	
x UTM coordinate	−0.0022	−0.6388	0.1331	1.000

Now to the key question: which hypothesis is confirmed? Which model best describes the geographic pattern of *Shorea robusta*? Since Models 1B (Hypothesis 1) and 2B (Hypothesis 2) are nested in Model 3 (Hypothesis 3), a likelihood ratio test can be used to determine, quantitatively, which hypothesis is supported (see Long, 1997, 93–97). First, a test was run to determine whether the data support Hypothesis 1 over Hypotheses 2 or 3. This tests whether the parameters for foraging and institutional effects (average household distance, elevation, and x UTM coordinate) are simultaneously zero. The results of the test confirm that Models 2 and 3 improve explanatory power (LR $x^2 = 13.37$, $df = 3$, $p < .01$). Some pattern related to optimal foraging exists in the plot data. Next, a test can be made between parameters associated with Hypotheses 2 and 3. The test is that the x UTM coordinate's coefficient is equal to zero. The results of this test confirm that the x UTM parameter also improves explanatory power (LR $x^2 = 6.51$, $df = 1$, $p < .05$). The coefficient in support of Pattern 3 is supported statistically.

Discussion

Substantive Findings

Earlier, three rival hypotheses were presented. Hypothesis 1 suggested that the distribution of *Shorea robusta* should be one of regrowth with no signs of human disturbance (figure 5.3). Hypothesis 2 suggested that the distribution of *Shorea robusta* should exhibit patterns of human disturbance and would be best predicted by optimal foraging theory (figure 5.4). Hypothesis 3 posed a different scenario, suggesting that the distribution of *Shorea robusta* would reflect optimal foraging altered by the geographic pattern of rule enforcement in the region (figure 5.5). The statistical results above support Hypothesis 3.

Of the several abiotic and biotic factors thought to influence the existence of *Shorea robusta* in this study site, only plot steepness is statistically significant. The soil A and B horizon depth provided little explanatory power. Other soil measures, such as soil color and texture (to get at concepts such as soil moisture), were applied in earlier analyses not shown and were also found to have little influence. This either means that other soil measures (such as soil nutrients) are required or that *Shorea robusta* can grow in a variety of soil conditions in this region.

The importance of optimal foraging parameters associated with Hypothesis 2 suggests that overharvesting in patterns anticipated by optimal foraging are occurring in the region. The rate of harvesting does appear to be outpacing the rate of *Shorea robusta* regrowth in areas within reasonable foraging access. While two parameters pick up some of this influence in Model 2B, it is elevation that proves to be a very important variable in determining the number of *Shorea robusta* trees in a plot. Because this tree species is known to grow in elevations much higher than the highest plot of 830 meters, the best explanation for why this variable is so important statistically is the altitudinal effort component of optimal foraging. It takes great effort to trek up to high elevations and bring harvested *Shorea robusta* trees down.

Hypothesis 3, the institutional-related influences of optimal foraging, is supported by the significance of the x UTM parameter in Model 3. Experience from the field led us to the hypothesis that more effective western forest monitoring and sanctioning, along with social hierarchical structure, lead villagers to forage further east than they would if the institutional structure were not there. The x UTM parameter identifies a depleting trend as one moves west to east toward the eastern side of the Latauli Forest and supports this contention. This suggests that the eastern side of the Latauli Forest in figure 5.2 has been subject to higher levels of *Shorea robusta* harvesting than the other forested areas.

The reader could argue that the identified trend might not be institutionally determined but rather be a result of some other still unidentified process. If one takes this view, the question then becomes, What better alternative explanations exist? In this analysis, every attempt has been made to control for abiotic, biotic, and optimal foraging parameters that explain *Shorea robusta* growth. I can think of only one other plausible explanation.

A rival explanation brings us back to the contention that differences in soil condition are not adequately measured in this study, and this may be the unknown process driving the trend. However, when considering this explanation, we run into the classic "Which came first—the chicken or egg?" dilemma. In their study of *Shorea robusta* forests only a few kilometers away from our study site, Burton, Shah, and Schreier (1989) report a relatively uniform soil base (e.g., C horizon) in the region. Their site is, in a biophysical sense, quite representative of this study area, and therefore it is unlikely that differences in C horizon cause the trend. The authors also report differences between degraded and nondegraded forest A and B soil horizons in the region. It may be inadequately measured soil-condition differences (such as nutrient content) in plot A and B horizons that explain this trend. But the Burton et al. (1989) study suggests that the opposite relationship exists. They report differences in soil condition measures that are related to where forests have been degraded by human action. In other words, their work suggests that where *Shorea robusta* is depleted, soil quality will be diminished and not the other way around. This suggests that missing soil-condition variables probably do not explain the identified trend. We are left with the institutional component providing the best explanation for the trend phenomenon.

Methodological Implications

While ecologists and biologists have made tremendous advances in the study of the spatial distribution of various plant species, to my knowledge, this is the first analysis of its kind that applies recent technological advances of DGPS, GIS, and a spatial statistical technique to this effort. This study provides an example of how the inclusion of a spatial influential variable in a regression model may assist in understanding the human dimensions of forest change when longitudinal data is nonexistent. The findings support the earlier claim that a plot level of analysis may reveal findings that would not be discovered at the forest level of analysis.

Moreover, this study may also be the first of its kind to apply an institutional analysis to the study of the distribution of a particular species over space. In any foraging setting, the first signs of forest depletion will be changes in geographic pattern of particularly important product species as humans base their decisions and actions on the attributes of the physiographic, institutional, and community norms related to use of that forested area. After taking into consideration the natural distribution of a particular species, and accounting for physiographic influences that encourage or discourage growth, the analyst can study the existing pattern to reveal human response to past and present institutional arrangements. In this case, such an analysis provides evidence that monitoring and social norms produce shifts in foraging patterns away from traditional optimal foraging. Such evidence could not be discovered with an analysis of aggregated cross-sectional forest-condition data.

The study raises the important problem of how we link spatial statistical analysis to the study of the geographic component of institutions. Rules and levels of their enforcement have geographic properties. In forest and watershed management, an important policy question is how the spatial configuration of institutions changes the spatial configuration of human action, which then leads to changes in the geographic properties of the natural resource. Granted, the X coordinate in this study is not the ideal parameter to capture the "institutional pressure" concept, but for statistical purposes in this study, it was the best parameter available. The advent of GIS, GPS, and spatial statistical procedures brings forward new methodological and measurement challenges to the social sciences. One challenge is to move beyond what has been done in this study by using better methods for understanding the spatial influence of rules on human behavior. A second important challenge we now face is how to measure the degree to which rules are followed and enforced that moves beyond a simple binary variable toward more of a continuous variable. Finally, a third and more problematic issue is how we can successfully inventory sets of potentially complex rule systems over broad landscapes. It may require empirical sampling of rules over geographic space.

Overcoming these challenges is important if we wish to understand the geographic ramifications of institutional or policy design. It is certain that when the rule and monitoring mechanisms were designed for this area of southern Nepal, it was not the intention to cause a harvesting shift into eastern areas deeper and higher into a watershed. Technological and statistical tools, such as the ones used here, provide the opportunity to devise methods that help us verify the performance and geographic implications of our institutional configurations in natural-resource management—something that has not been easily conducted before. I hope this work inspires others to explore the utility of such methods.

Acknowledgments

This chapter is forthcoming in *Environmental Monitoring and Assessment*. Reprint permission from Kluwer Academic Publishers is appreciated. I am also very appreciative of the support received from the Ford Foundation, Dr. John Ambler, and Dr. Ujjiwal Pradhan, as well as the financial and intellectual support provided by Elinor Ostrom at the Workshop in Political Theory and Policy Analysis, Indiana University, Bloomington. The Forests, Trees, and People Programme of the United Nations Food and Agriculture Organization was also quite helpful through its support of the IFRI research program. Special thanks go to Rajendra Shrestha, Bharat Mani Sharma, Mukunda Karmacharya, Vaskar Thapa, and Sudil Gopal Acharya for their efforts in data collection. I am also grateful to K. N. Pandit, K. R. Adhikari, A. K. Shukla, Ganesh Shivakoti, and the Institute of Agriculture and Animal Science in Rampur, Chitwan, for their assistance in the field. I am indebted to Dusty Becker, Erling Berge, Brenda Bushouse, Clark Gibson, Joby Jerrells, Robert and Joanne Schweik, John Williams, and five anonymous reviewers for extremely helpful comments on earlier drafts. Thanks as well to Robin Humphrey and Julie England, who provide ongoing database and technical support with amazingly good cheer, and to Patty Dalecki, for her outstanding editorial assistance. Most of all, I am deeply grateful for the openness and warmth given by residents of the Shaktikhor VDC in Chitwan, Nepal, during our days in the field.

Notes

1. Villager names used here differ slightly from the names utilized in another study of this general region (see Schweik, Adhikari, and Pandit, 1997). For consistency purposes in this analysis, village names correspond to the names printed on His Majesty's Government scanned topographic map displayed in Figure 5.2. The Schweik, Adhikari, and Pandit study, on the other hand, provides village names as reported by the villagers themselves.

2. For example, Smith (1983, 632) describes the role that "exclusive control" plays in the conservation of natural resources. Feit (1973) describes rotational hunting by the Waswanipi Cree people as a method for controlling the size of animal population.

3. *Density* provides a measure of the number of species present in a forest. It is determined by counting the number of individual species and then dividing this by the total area of plots. *Frequency* provides a measure of how widely a species is distributed within a forested area. It is calculated by taking the number of plots in which a species occurs and dividing this by the number of plots sampled. *Dominance* provides a measure of the standing biomass a particular species contributes

to a forest composition. Dominance is calculated by taking the total basal area of a species and dividing it by the area sampled. Finally, the *importance values of each species* reports the summation of the *relative* density, dominance, and frequency together divided by three.

4. The question arises whether the dependent variable should be treated as a truncated or nontruncated variable and whether a negative binomial or a zero-inflated negative binomial is required (see Long, 1997, chap. 8). The dependent variable is a count of all trees with DBH > 10 cm. The negative-binomial regression model assumes that each plot has a positive probability of producing a tree > 10 DBH. Given that these are *Shorea robusta* climax forests with *Shorea robusta* trees appearing everywhere and that we are accounting for abiotic factors in the model, this is a reasonable assumption.

5. Model 1B is surprisingly simple. But recall that many other abiotic and biotic factors (such as elevation, insect or animal damage, and competing biomass) were considered for inclusion in the model and rejected.

References

Aldhous, P. 1993. "Tropical Deforestation: Not Just a Problem in Amazonia." *Science* 259: 1,390.

Angelsen, A. 1995. "Shifting Cultivation and 'Deforestation': A Study from Indonesia." *World Development* 23(10): 1,713–29.

Ascher, W. 1995. *Communities and Sustainable Forestry in Developing Countries.* San Francisco: ICS Press.

Bruijnzeel, L. A., and C. N. Bremmer. 1989. "Highland-Lowland Interactions in the Ganges Brahmaputra River Basin: A Review of Published Literature." ICIMOD Occasional Paper No. 11. ICIMOD, Kathmandu, Nepal.

Burton, S., P. B. Shah, and H. Schreier. 1989. "Soil Degradation from Converting Forest Land into Agriculture in the Chitwan District of Nepal." *Mountain Research and Development* 9(4): 393–404.

Feit, H. A. 1973. "The Ethno-ecology of the Waswanipi Cree, or How Hunters Can Handle Their Resources." In *Cultural Ecology,* ed. B. Cox. Toronto: McClelland and Stewart.

Hayden, B. 1981. "Subsistence and Ecological Adaptations of Modern Hunter-Gatherers." In *Omnivorous Primates,* ed. R. S. O. Harding and G. Teleki. New York: Columbia University Press.

His Majesty's Government, Nepal. 1994. 1:25,000 Scale Maps # 2784 07A and 2784 07B. Kathmandu, Nepal: Survey Department.

International Forestry Resources and Institutions (IFRI) Research Program. 1994. *IFRI Data Collection Forms of the Shaktikhor Site, Nepal.* Bloomington: Indiana University, Workshop in Political Theory and Policy Analysis.

Keller, M., D. A. Clark, D. B. Clark, A. M. Weitz, and E. Veldkamp. 1996. "If a Tree Falls in the Forest . . ." *Science* 273: 201.

King, G. 1988. "Statistical Models for Political Science Event Counts: Bias in Conventional Procedures and Evidence for the Exponential Poisson Regression Model." *American Journal of Political Science* 32: 838–63.

———. 1989a. "Event Count Models for International Relations: Generalizations and Applications." *International Studies Quarterly* 33: 123–47.

———. 1989b. "Variance Specification in Event Count Models: From Restrictive Assumptions to a Generalized Estimator." *American Journal of Political Science* 33: 762–84.

Leamer, E. E. 1983. "Let's Take the Con Out of Econometrics." *American Economic Review* 73(1): 31–43.

———. 1985. "Sensitivity Analysis Would Help." *American Economic Review* 75(3): 308–13.

Long, J. S. 1997. *Regression Models for Categorical and Limited Dependent Variables*. Thousand Oaks, CA: Sage.

Lovejoy, T. E. 1980. "A Projection of Species Extinctions." In *The Global 2000 Report to the President: Entering the Twenty-first Century*, ed. G. O. Barney. Washington, DC: Council on Environmental Quality, U.S. Government Printing Office.

Ludwig, J. A., and J. F. Reynolds. 1988. *Statistical Ecology*. New York: Wiley.

McKean, Margaret A. 1992. "Management of Traditional Common Lands (*Iriaichi*) in Japan." In *Making the Commons Work: Theory, Practice, and Policy*, ed. D. W. Bromley et al. San Francisco: ICS Press.

Metz, J. 1990. "Forest-Product Use in Upland Nepal." *The Geographic Review* 80(3): 279–87.

Morrow, C. E., and R. W. Hull. 1996. "Donor-Initiated Common Pool Resource Institutions: The Case of the Yanesha Forestry Cooperative." *World Development* 24(10): 1,641–57.

Myers, N. 1988. "Tropical Forests and Their Species: Going, Going . . ." In *Biodiversity*, ed. E. O. Wilson and F. M. Peter. Washington, DC: National Academy Press.

Norton, B. J., ed. 1986. *The Preservation of Species*. Princeton, NJ: Princeton University Press.

Ostrom, E. 1990. *Governing the Commons: The Evolution of Institutions for Collective Action*. New York: Cambridge University Press.

Ostrom, E., and M. B. Wertime. 1994. "IFRI Research Strategy." Working paper. Workshop in Political Theory and Policy Analysis, Indiana University, Bloomington.

Pace, S., et al. 1995. *The Global Positioning System: Assessing National Policies*. Santa Monica, CA: Rand.

Pickett, S. T. A., and M. L. Candenasso. 1995. "Landscape Ecology: Spatial Heterogeneity in Ecological Systems." *Science* 269: 331–34.

Reid, W. V., and K. R. Miller. 1989. *Keeping Options Alive: The Scientific Basis for Conserving Biodiversity*. Washington, DC: World Resources Institute.

Repetto, R. 1988. *The Forest for the Trees? Government Policies and the Misuse of Forest Resources.* Washington, DC: World Resources Institute.

Richards, J. F., and R. P. Tucker, eds. 1988. *World Deforestation in the Twentieth Century.* Durham, NC: Duke University Press.

Ruser, J. W. 1991. "Workers' Compensation and Occupational Injuries and Illnesses." *Journal of Labor Economics* 9(4): 325–50.

Schreier, H., S. Brown, M. Schmidt, P. Shah, Nakarmi G. Shrestha, K. Subba, and S. Wymann. 1994. "Gaining Forests but Losing Ground: A GIS Evaluation in a Himalayan Watershed." *Environmental Management* 18(1): 139–50.

Schweik, C. 1997. "The Spatial Analysis of Natural Resources in East Chitwan, Nepal: Conceptual Issues and a Multi-scale Research Program." In *People and Participation in Sustainable Development: Understanding the Dynamics of Natural Resource Systems,* ed. G. Shivakoti et al., 219–34. Proceedings of an International Conference held at the Institute of Agriculture and Animal Science, Rampur, Chitwan, Nepal, 17–21 March 1996.

Schweik, C. M., K. Adhikari, and K. N. Pandit. 1997. "Land-Cover Change and Forest Institutions: A Comparison of Two Sub-basins in the Southern Siwalik Hills of Nepal." *Mountain Research and Development* 17(2): 99–116.

Shrestha, R. 1996. IFRI Team leader, Personal communication (June).

Smith, E. A. 1983. "Anthropological Applications of Optimal Foraging Theory: A Critical Review." *Current Anthropology* 24(5): 625–52.

Spurr, S. H., and B. M. Barnes. 1992. *Forest Ecology.* Malabar, FL: Krieger.

Storrs, A., and J. Storrs. 1990. *Trees and Shrubs of Nepal and the Himalayas.* Kathmandu: Pilgrims Book House.

Task Force on Global Biodiversity, Committee on International Science. 1989. *Loss of Biological Diversity: A Global Crisis Requiring International Solutions.* Washington, DC: National Science Board.

Thomson, J. T., D. Feeny, and R. J. Oakerson. 1992. "Institutional Dynamics: The Evolution and Dissolution of Common-Property Resource Management." In *Making the Commons Work: Theory, Practice, and Policy,* ed. D. W. Bromley et al. San Francisco: ICS Press.

Umans, L. 1993. "The Unsustainable Flow of Himalayan Fir Timber." *Mountain Research and Development* 13(1): 73–88.

Winterhalder, B. 1993. "Work, Resources and Population in Foraging Societies." *Man* 28: 321–40.

6

A Lack of Institutional Demand: Why a Strong Local Community in Western Ecuador Fails to Protect Its Forest

Clark C. Gibson and C. Dustin Becker

Introduction

Given the disappointing results of natural-resource conservation policy in developing countries over the last three decades, scholars and practitioners have shifted their focus away from state-centered policies toward solutions at the local level (Ostrom, 1990; Hecht and Cockburn, 1990; Marks, 1984; Blockhus et al., 1992; Poffenberger, 1990; Bromley et al., 1992; McCay and Acheson, 1987; UNFAO, 1990; Ascher, 1995). While these authors offer different lists of the conditions believed necessary for successful resource management by local people, most analyses include three fundamental requirements. First, individuals from local communities must highly value a natural resource to have the incentive to manage it sustainably. Second, property rights must be devolved to those individuals who use the resource to allow them to benefit from its management. Third, these individuals at the local level must also have the ability to create microinstitutions to regulate the use of the resource. Although various scholars and practitioners may add other conditions they see as important, most agree that some form of these three—locals' valuation, ownership, and institutions—are central to successful natural resource management.

In the *comuna* of Loma Alta in western Ecuador, these three conditions initially appear to be met. Residents of Loma Alta consider their 1,650 hectares of tropical moist forest important for its products such as timber to sell, building materials, and game. Comuna members enjoy well-defined and secure property rights to their land, allowing individuals to make capital improvements to their plots, rent their lands to others, and transfer their holdings to family members through inheritance. Finally, Loma Alta boasts a strong history of crafting local institutions to deal with community concerns. The community has successfully crafted institutional arrangements dealing with the provision of goods such as schools, health clinics, and wells, as well as electoral institutions that allow each comuna member a voice in the administrative proceedings and the selection of their leaders. The central government has recognized the comuna as the legitimate form of local government since the Law of the Comunas passed in 1936.

Despite their positive valuation of the tropical forest, their relatively secure property rights to land, and their rich history of crafting micro-institutions, the members of the Loma Alta community have not created microinstitutions to regulate the use of their tropical forest. Few local rules exist about the removal of forest products, the cutting of timber, the hunting of game, or the clearing of land. Although parts of the forest appear to be relatively healthy, over a third of the forest has been decimated by the exploitation of timber and the expansion of agricultural and pasture lands.

Some of the explanation for the forest's depletion can be found in the type of property rights the comuna has allocated to different parts of the forest. The one-third that is most exploited is the comuna's "forest reserve," which has not been allocated to individual comuna members. This section's overuse conforms to outcomes predicted by well-known theories regarding open-access, common-pool resources. The other two-thirds of the comuna's forest has been allocated and is in relatively better condition. And yet, property rights alone do not explain the spatial variance of the forest's condition within the allocated areas. Some individuals with plots in the forest appear to cut selectively their plots, generating stands of secondary growth. Others, however, pursue plantation agriculture or cattle raising, motivating them to clear the forest to expand their holdings. The result of this complex pattern of property rights and activities is a starkly patchy forest: nearly treeless areas are contiguous with sections of dense secondary growth containing a wide diversity of species, some endemic to the region.

This chapter seeks to explain why the members of Loma Alta have not created microinstitutions to protect and manage their forest. Unlike so many local communities in the developing world, Loma Alta does possess those institutional features considered necessary for the successful conservation of natural resources; yet it, too, has failed to create rules to protect its forest. We argue that the lack of institutions regarding the forest requires an understanding of the forest's many user groups, the forest products they value, and their property rights to these products. We find that the pattern of incentives confronting Loma Alta's multiple forest users discourages the creation of institutions to govern forest use, despite the comuna's strong institutional assets. Comuna members prize the immediate exploitation of certain forest products and do not recognize the critical public goods produced by the forest, especially watershed and climatic services. Only when comuna members substantially value the benefits of these public goods and overcome the collective-action problem of institutional supply will a local-level institution regulating the Loma Alta Forest be created.

We collected our data using the methods of the International Forestry Resources and Institutions (IFRI) research program. The IFRI program is a pioneering effort to study forests and their use by collecting and analyzing both social and biological data at the micro level. A central hypothesis of this program is that institutions significantly affect the use and condition of forests (see the appendix to this volume).

This chapter has six parts. In the first part, we briefly review some of the core assertions made by scholars and practitioners regarding the supply of microinstitutions that govern natural-resource use at the local level. The next part introduces the comuna of Loma Alta, reviewing its institutional history, decision-making structures, and property-rights institutions. In the third part, we present the biological data collected in the Loma Alta Forest. These data indicate that much of the forest is in relatively good shape, while some parts exhibit tremendous overuse. In an attempt to explain the variation of forest condition, we investigate the users of the forest, their use patterns, and the rules that influence their behavior in the fourth part. We show that the groups that comprise the greatest contemporary threats to the forest's condition are comuna members and outsiders using unallocated land, and members who convert forest land to plantation agriculture. We present an analysis of these use patterns in the next part, attempting to derive an explanation for the lack of institutional supply from the incentives of user groups. Creating institutions to manage natural resources is costly; such costs are increased by the multiuser, multiproduct nature of forests. In Loma Alta, individuals do not value the public goods generated by the forest, and the different streams of private benefits that accrue to individuals are not sufficient to motivate them to create rules to regulate forest use. In fact, the three most important user groups in Loma Alta—farmers, woodcutters, and outsiders—would experience significant losses in the short run if an institution restricted their use of the forest. We conclude the chapter by discussing how the pattern of user-group behavior may be changed in an effort to prevent the Loma Alta Forest from being completely depleted.

Natural-Resource Management and the Local Level

A growing number of scholars and practitioners recognize the crucial role played by local people in natural-resource management (Ostrom, 1990; Hecht and Cockburn, 1990; Marks, 1984; Blockhus et al., 1992; Poffenberger, 1990; Bromley et al., 1992; McCay and Acheson, 1987; UNFAO, 1990; Ascher, 1995; Agrawal, chapter 3 this volume). They argue that policies emanating from central governments generally give local communities few rights over the natural resources with which they live. Without legal claims to the stock or flow of benefits from these resources, locals have little to gain from protecting them or using them sustainably. Such conditions generate incentive structures that encourage individuals to "poach" natural resources and discourage them from constructing or maintaining rules or institutions at the local level to regulate their resource use (Gibson and Marks, 1995). Because many governments lack the resources necessary to monitor and enforce their natural-resource policies, this pattern of incentives often results in overexploited resources (Becker, Banana, and Gombya-Ssembajjwe, 1995).

Critics of exclusionary government policies assert that sustainable policies must include those individuals that live with the natural resource. Many conditions for successful local-level management have been put forward. Most writers, however, include three requirements: (1) locals must value the resource, (2) they must possess some property rights to the resource, and (3) they must construct local-level institutions that control the use of the resource (Bromley et al., 1992; McCay and Acheson, 1987; Ostrom, 1990; McKean, chapter 2 this volume). The reason for the first condition is clear: unless locals place sufficient value on the resource, they have no reason to incur costs to protect or conserve it. While this condition appears trivial, many scholars and public policymakers routinely ignore it and think that individuals will somehow conserve resources for some national or global good. Most practitioners, however, have come to realize that people must perceive some individual net gains from managing a resource to agree to constrain their short-term use of it.

The second condition of successful local management highlights the importance of property-rights arrangements. While debate surrounds exactly which bundle of property rights is most efficient for the sustainable use of natural resources, considerable agreement exists that locals should have some stake in the resource relating to access, use, and the exclusion of others (McKean, chapter 2 this volume; Demsetz, 1967; Libecap, 1989; North, 1990; Ascher, 1995). Such rights allow locals to control the benefits and costs of a resource and thus may offer a reason for people to manage it for the long term (Schlager and Ostrom, 1993).

Finally, scholars and practitioners often assert the need for local-level institutions in natural-resource management schemes (Ostrom, 1990; Marks, 1984; Bromley et al., 1992). When compared to central government institutions, local institutional arrangements are considered better at providing, *inter alia*, rules related to access, harvesting, and management; fora that can respond to conflict quickly and cheaply; and monitoring and sanctioning methods that are efficacious. Further, locals are more likely to create such institutions if their community enjoys a history of rule making together, since the costs, benefits, and techniques of institution building will be well known to the participants.

These three general conditions are by no means exhaustive of the requirements authors assert are important to the construction of successful natural-resource management institutions. Others include sufficiently small boundaries for the resource to be managed, a relatively small number of users, users who live near to the resource, users who are not strongly divided by cultural or ethnic differences, and users who perceive the rights system to be relatively fair. The case of the Loma Alta comuna in western Ecuador not only meets the three general criteria presented but fulfills almost all of the preconditions that scholars and practitioners consider important.

The Social and Physical Assets of Loma Alta

The Loma Alta comuna is a community of approximately 2,000 people who share property rights to 6,842 hectares of land in western Ecuador (see figure 6.1). The comuna members are distributed among four settlements—Loma Alta, La Union, La Ponga, and El Suspiro (see figure 6.2). Current residents recount how the settlements were established at the turn of the century by five families moving from more populated towns of the east and southwest who were seeking better opportunities for themselves; they especially sought to acquire land for agriculture. Small numbers of peoples were indigenous to the region and had established land-tenure patterns roughly based on the watersheds of the Chongon Colonche mountain range. The newer settlements continued this centuries-old pattern, as well as the linkages with small towns on the coast to supplement their household needs. These early settlers survived through subsistence farming and selling charcoal, timber, and straw hats to townsfolk.

In response to the actions taken by several coastal municipalities that were selling large tracts of land to urban dwellers during a period of land speculation, the central government passed the Law of the Comunas in 1936. This law formalized and augmented much of the traditional land-tenure arrangements already found in the area. Individuals can petition a comuna to be a member when they reach the age of 18. Members pay an annual tax that is used to provide and maintain certain public goods in the comuna (such as the health clinic and roads). Governing the comuna occurs through two institutions. The comuna chooses a *cabildo* (council) each year in democratic elections decided by majority rule. Five officers comprise the cabildo—president, vice-president, treasurer, secretary, and legal advisor—and are responsible for the comuna's daily management. The cabildo officers also chair the monthly

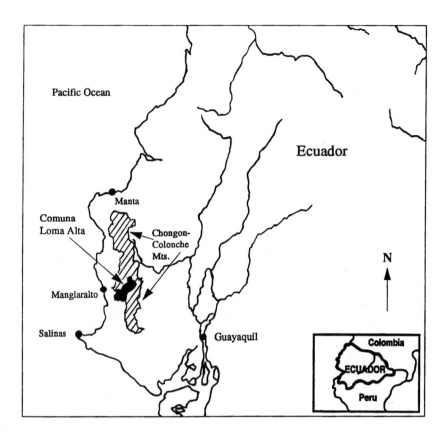

Figure 6.1
Map of western Ecuador

asamblea (community meeting) at which all comuna members make decisions collectively through majority votes. Members are expected to attend regularly and can be punished if they are absent from the asamblea.[1] Members also frequently serve on various comuna committees (existing committees include child care, education, sanitation, and reforestation).

The most critical power of the comuna is its control over land. The 1936 law stipulates that the comuna as a whole owns the land and can allocate it to members for their use. In Loma Alta, a member must petition the comuna for land; asambleas usually grant most requests for plots less than 15 ha (although many members possess more than one plot). Several

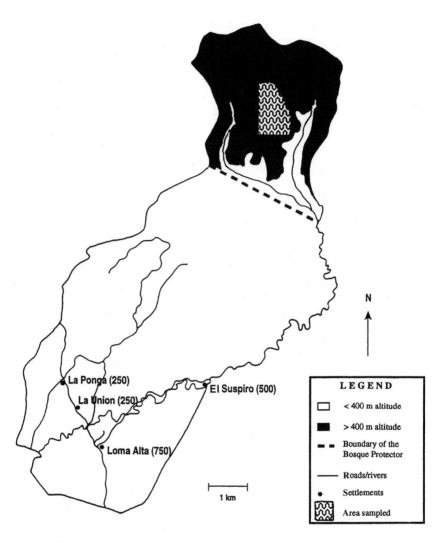

Figure 6.2
Map of the Loma Alta comuna and its Bosque Protector

rules constrain members' rights to their land. First, the comuna allocates land with the understanding that it must be used; plots left unused are subject to confiscation by the comuna. In practice, however, the interpretation of *use* is quite broad in Loma Alta: the comuna considers renting a plot to be a bona fide use, as well as keeping a field fallow for the regeneration of trees. No current member of Loma Alta recalls an incident in which the comuna has reclaimed land previously allocated. Second, an individual cannot sell his or her land to an outsider without the comuna's approval (by majority vote at an asamblea). To date, no land has been sold by comuna members.[2] Third, a member cannot rent land to anyone without comuna permission. Members, however, routinely flout this rule, renting land to other members without informing the comuna. Fourth, on a member's death, land returns to the comuna to be reallocated. If any improvement to the land had been made by the deceased member, however, the comuna is required to compensate family members at the market price of the improvement(s). In practice, this compensation clause acts to promote inheritance. Since the comuna rarely has the money to recompense family members for improvements, sons and sons-in-law invariably receive their fathers' plots. No one in Loma Alta remembers an example of property reverting to the comuna after a member's death. Still, sons and sons-in-law often make official "requests" to the comuna for their fathers' land so as ensure this inheritance.

Comuna members respect each other's land boundaries. When the comuna decides to allocate a plot, a cabildo officer (or representative appointed by an officer) will travel to the plot site with the prospective user. The official, prospective owner, and neighbors agree on the new boundaries, which can be either part of the natural landscape (such as a river or a ridge top) or constructed (with rocks, planted trees, and so on). This system appears to work relatively well, as comuna members and officials consider boundary disputes among comuna members to be rare. Incursions by individuals outside of the comuna, however, do occur. The most egregious example of such incursion occurs in the comuna's tropical premontane humid forest, which we discuss below.

Several consequences flow from this system of property rights to land. First, members hold considerable rights to their property: they are not restricted in their use of land, face few impediments when renting it, and can inherit land from family members. While they cannot sell their plots outright, they possess enough incentive to make considerable capital investments in the land, as evidenced by the number of houses built, wells sunk, fences constructed, trees planted, and irrigation trenches dug in Loma Alta. Thus, although the entire comuna system possesses some "communal" attributes, those allocated land within Loma Alta enjoy strong private rights to their property. As we show below, these rights critically affect the use and condition of the Loma Alta Forest.

Loma Alta's "Protective Forest"

In 1986, Loma Alta sought assistance from Ecuador's central government to have its upland territory protected from encroachments made by members of a neighboring comuna. The area lies approximately 8 kilometers from El Suspiro, the nearest settlement, requiring three to four hours of travel time to reach (local residents travel on foot, mule, and horse). By 1987, the Ministry of Agriculture had demarcated the northern 1,650 ha of the comuna and declared it a *Bosque Protector* (Protective Forest; hereafter "the forest") (see figure 6.2).[3]

The forest exists on steep hills ranging in altitude from 200 to 830 meters. Along that gradient, vegetation changes from predominately tropical dry forest to a premontane humid "fog forest."[4] Much of the moisture required to support the moist forest tree species comes from the *garua* or fog season that lasts from July through November. Fog interception supports trees typically found in wetter regions of Ecuador and enables abundant populations of epiphytes to grow in the forest.[5]

Ecologists divide the forest into two ecological zones. Those parts above 400 m are dominated by premontane humid forest (Fundacion Natura, 1992). At elevations below 400 m, the forest shifts to dry forest, which contains more deciduous species. The transition between these two ecological zones is not abrupt. While the moisture of the forest increases from lower to higher elevations, the type of crops planted by Loma Alta's farmers does not vary much at elevations above 300 m—which includes almost all of the forest lands except river valleys.

For this analysis, another useful division of the forest follows the different property rights assigned to its parts by the comuna. Much of the

northwest portion of the forest has not been allocated to individual comuna members, who call this area the *comuna reserve*. In this area, which
we estimate to cover approximately 600 ha, all comuna members may
extract resources.[6] In the remaining 1,050 ha of the forest, the comuna
has allocated plots to individuals, who enjoy the bundle of rights discussed earlier.

The condition of the forest can be partly explained by these two different sets of property rights. The open-access nature of the comuna reserve
has led to its severe degradation. An aerial photograph taken in August
1986 shows deforestation along the entire northern and western edges
of the comuna reserve. At that time, about 50 ha had been converted to
pasture, and another 50 ha had been cleared and cultivated. By August
1995, the pasture in the comuna reserve had been extended to cover approximately 350 ha, and extensive timber harvesting had taken place on
the rest. We estimate that as a whole, users' intensive exploitation of the
comuna reserve has led to the removal of 75 percent of the area's forest
cover.

In contrast, in the part of the forest that has been allocated to individuals, the forest is less depleted overall. However, the allocated areas display
considerable variation over forest condition both within and between
parcels. Such variance results from the different types of activities that
landholders pursue on their plots. Those comuna members who are engaged in agriculture value the lands of the forest because of the increased
humidity in the area. These farmers are slowly intensifying agricultural
practices in the forest in response to the drying trend found at lower
elevations.[7] Two major stream beds provide landholders easy access to
this area during the dry season, as well as water for their crops from
February through March.

Most of the comuna members with plots in the forest have responded
to these favorable conditions by planting *paja toquilla* (*Carludovica palmate*), the leaves of which are sold to the makers of panama hats. Farmers' holdings vary from approximately 5 to 12 ha, with 1 to 3 ha
established as paja toquilla plantations. On these plantations, the forest
is cleared of forest trees, burned, and planted with paja seedlings. In some
of the remaining areas of their holdings, farmers plant crops such as citrus
trees, plantain, tagua, banana, and coffee.[8] In between and among these

crops can be found stands of secondary growth forest, although we estimate that only a handful of trees with a diameter at breast height (DBH)
of more than 25 centimeters remain.

Woodcutters also own plots within the forest's boundaries, and it was
on these plots that we conducted our most in-depth biological analyses.[9]
Interestingly, the condition of the forest on woodcutters' plots is generally
quite good. In 30 plots of 300 square meters each, randomly distributed
over 200 ha of landholdings, only two plots had any recent (within the
last five years) evidence of timber harvesting.[10] Additionally, we found
no cases of current conversion to agriculture or pasture in these forest
landholdings.

For some timber species in this part of the forest, it is obvious that
sustainable harvesting has not been the norm, and that the resource has
been depleted. For example, only four of the 493 trees measured were
the extremely valuable guayacan (*Tabebuia chrysantha*), and no saplings
or seedlings of this species were recorded. Still, per hectare, our sampling
found 30 preferred timber trees with diameters above 25 cm, and regeneration was occurring for many of these. Using this number as an estimator
for the entire 200 ha sampled in our study, about 5,962 timber trees of
harvestable size currently exist, or 5.7 percent of the trees (523 trees per
ha × 200 ha = 104,600) we estimate to remain.

The size class distribution and the density of the current fog forest stand
reflects the harvest of older, larger trees in the past. The mean DBH of
trees with a DBH above 10 cm is only 21.8 ± 16.34 cm ($N = 492$ trees),
indicating a young or secondary forest structure.[11] Primary tropical forests are surprisingly consistent in proportions of stems of a particular size
(age) (Richards, 1975). As shown in table 6.1, our data from the Loma
Alta Forest deviates from the primary forest pattern in an expected way.
In the secondary forest of Loma Alta, there are more small trees in the
10 to 20 cm category, and fewer large, older trees, explaining the low
average stem diameter.

The density and diversity of mature trees with DBH greater than 10 cm
are shown in table 6.2. These structural and community features are consistent with expectations for a normal regenerating secondary forest. Typical of selectively harvested forests, the Loma Alta Forest has a high
number of mature stems per hectare (523) and has patchy distributions

Table 6.1
Tree-size classes in tropical primary forests versus Loma Alta's secondary forest

Stem Class (diameter at breast height in centimeters)	Primary Forest (percent ±S.D.)[a]	Loma Alta (percent)
10–19.9	44 ± 4	0
20–20.9	28 ± 2	22
30–30.9	18 ± 2	8
40 and above	12 ± 2	10

Note: The distribution is statistically different.
a. N = 7 primary forests—3 South America, 2 Africa, 2 Asia (Richards, 1975, 230).

of pioneer genera such as *Cercropia, Inga,* and *Geonoma.* Gaps created by the harvesting of the large timber trees are being filled by these fast-growing soft-wood and palm species. Species-abundance patterns are normal for tropical second-growth forests with four or five dominant species, four or five subdominants, and a long list of less common species.

These findings are hardly what one would expect if the Loma Alta community had used its entire forest as an open-access resource.[12] Neither are they consistent with what we would expect if Loma Alta had constructed institutions to manage their forest resources, purposefully maintaining the 1,650 ha of protective forest. Rather, variation of forest cover in the Loma Alta Forest reflects the practices of different user groups operating under different sets of incentives.

Users, User Rules, and Use Patterns

Different subsets of comuna and noncomuna members value the assets of their forest for different reasons. In this part, we examine the six most important user groups of the Loma Alta Forest: hunters, outsiders, wood users, commercial timber dealers, farmers, and woodcutters. Some, but not all, of the individuals of these groups overlap. The resultant pattern of users, products, and preferences helps explain the variance of the forest's current condition.

Hunters comprise one important group using the Loma Alta Forest. While populations of wild game in the forest have declined over the years,

Table 6.2
Diversity and density of trees (DBH > 10 cm) in Loma Alta's fog forest, Ecuador

Taxonomic Information (local names)	Stems (per hectare)	Estimated Percent of Trees
I. Preferred timber	80	15.3
Beilschmiedia spp. (Maria)	36	6.8
Ocotea spp. (jigua)	20	3.8
Cordia spp. (tutumbe)	10	1.9
Guarea spp. (chicoria)	8	1.5
Tabebuia chrysantha (guayacan)	4	.8
spp.? (figueroa, cedro)	2	.4
II. Taxon with more than 5 stems per hectare	345	66.0
Gleospermum sp. (guayaba de monte)	76	14.5
Quararibea grandifolia (molinillo)	65	12.4
sp? (morocho)	32	6.1
Geonoma sp. (palma)	25	4.8
Cecropia spp.	23	4.5
Chrysophyllum sp. (mangillo)	19	3.6
Grias sp. (huevo de chivo)	16	3.0
Mapuira sp. (camaron)	15	2.8
Inga spp. (guaba de bejuco)	13	2.5
Pentagonia sp. (palo de murcielago)	13	2.5
sp? (pepito colorado)	10	1.9
Turpinia occidentalis	8	1.5
Ficus spp. (mono, cauchillo)	8	1.5
Rheedia sp. (amarillo)	6	1.2
sp? (miguelillo)	6	1.2
Phylotacea dioica (yuca de raton)	5	.9
Randia sp. (canafito)	5	.9
III. Taxon with less than 5 stems per hectare	54	10.3
Prunus subcorymbosa (mamecillo)	4	
Sapium utile	4	
Zanthoxylum sp.	4	
Mauria sp. (mulato)	4	
Mollinedia sp. (cafe de monte)	4	
Pourouma sp.	4	
sp? (bijama)	3	
sp? (tabaquillo)	3	
Annona sp.	3	
Brosmium sp.	3	

Table 6.2 (continued)

Taxonomic Information (local names)	Stems (per hectare)	Estimated Percent of Trees
Piper squamulosum	2	
sp? (anona de monte)	2	
Ardisis sp.	2	
Bactris sp.	2	
Gutiferae sp.	2	
Phytelephas aequatorialis (tagua)	2	
Miconia sp.	2	
Psychotria sp.	2	
Tabernaemontana sp.	1	
Trema micrantha	1	
IV. Unidentified trees	44	8.4

enough paca (*Agouti paca*), guatusa (*Dasyprocta punctata*), white-tailed deer (*Odocoilus virginiamus*), and red brocket deer (*Mazama americana*) exist to encourage locals to make the trek to the forest to obtain meat. Comuna members seem to prefer the taste of game to that of domesticated animals (locals raise cattle, pigs, chickens, turkeys, ducks, and goats), but the price of game meat does not reflect this because game is not significantly more expensive. A trade in game meat does exist, but it is small and localized. Hunting is clearly secondary to residents' other activities. While it provides some additional protein to diets, it is not a critical supplement.

While the comuna has not established any formal rules regarding hunting within the forest, several norms appear to be respected by the hunters. First, individuals hunt alone or in small groups rarely exceeding four people; larger hunting parties are considered inappropriate. Second, hunters dislike spending nights in the forest, and so hunting trips of more than two days rarely occur. Third, comuna members disapprove of hunting for commercial gain. Those that do hunt generally eat what they kill, only occasionally selling small, extra portions to other comuna members.

Outsiders invading the forest constitute another significant user of the comuna's forest. The most important invader is a relatively wealthy, cattle-raising family living in a neighboring comuna (Dos Mangas). The family's employees have cut down the trees and burned the scrub on approximately 400 ha in the northern section of the comuna reserve.[13] The area cleared corresponds to several of the denuded patches evident on the 1986 aerial photograph, and our own efforts at ground-truthing discovered that the fenced pasture has been extended to an even greater area. While the comuna has made some efforts to prevent this incursion— through such means as having the forest declared protected, cutting the wire fences that the family's employees erect, attempting to use the courts, and confiscating lumber taken by the family from that plot—Loma Alta has few efficacious enforcement mechanisms to protect its comuna reserve.

A third important group, which includes most of the comuna's residents, uses the timber of the forest for construction. While some residents construct their homes and shops with concrete block or stone (especially in the town of Loma Alta, which is the most commercial settlement of the comuna), most of the people living in the settlements of El Suspiro, La Ponga, and La Union use the hardwoods and bamboo gleaned from the forest to build their homes, fences, animal pens, and small stores. Locals prize guayacan (*Tabebuia chrysantha*) for cross beams, maria (*Beilschmiedia spp.*) for stilts, and jigua (*Ocotea spp.*) for floor planking. Bamboo (*Guadua spp.*) is used for internal and external walls and is also an important fence-building material in all four of Loma Alta's settlements.

Individuals confront several choices in their efforts to obtain wood for construction. They can contract with landholders whose plots have the desired timber. They can also travel to a neighboring comuna to either poach or contract for timber. They can travel to the comuna's reserve— where land has not been allocated to any individual—to cut trees. Finally, they can contract with a woodcutter who will, in turn, cut the timber from the unowned reserve, negotiate with a landowner, or cut from a neighboring comuna. Comuna residents believe that the vast majority of wood currently taken comes from the comuna reserve. The constant use of this open-access area has resulted in local complaints about the increasing difficulty of finding the most-desired species for home building.

Individuals involved in the commercial timber business comprise another significant group of forest users—arguably the most critical user group when considering the forest's current condition. Timber was needed to build the coastal towns in the region (for example, La Libertad, Barcelona, Manglaralto, and Santa Elena). As a result, from 1940 through 1960, commercial timber interests cut extensively from the entire Chongon Colonche range. Loma Alta residents claim that these outsiders continued to cut in their forest to supply the towns with wood; only within the last decade has the commercial activity tapered off. Typically, outside merchants would arrive with trucks and either contract with comuna members who held land in the forest, contract with members who were woodcutters, or try to cut wood in areas held by the comuna as a whole to avoid payment.

Few rules appear to have limited the activities of the commercial timber industry. The comuna did make a small attempt to capture the benefits from this lucrative industry by imposing a tax on wood leaving their territory. However, since the tax was nominal and loosely enforced, it did nothing to restrain the cutting of trees. The intensity of this business has decreased noticeably with the concomitant reduction of commercially valuable timber. Currently, only a few trucks come to Loma Alta with the intent to transport timber out of the comuna. The lack of valuable species and large trees in most of the forest is in part attributable to the extensive cutting of previous generations.

Since large-scale commercial timbering has declined, the user group comprised of the approximately 25 comuna members who have been allocated plots within the forest has the most significant effect on the condition of the Loma Alta Forest.[14] Most of these landholders have cleared their plots to cultivate paja toquilla. Paja has been farmed in the area for at least the last 100 years. Its importance has grown over the past two decades due to the increasing demand for panama hats and the decline of its cash crop rivals—coffee and tagua. Comuna members have enjoyed a consistently growing demand for their paja leaves over the past generation; presently, it is the most valuable agricultural commodity in the comuna, and all of Loma Alta's farmers wish to expand their holdings. Two factors constrain the expansion of paja farming. First, paja toquilla requires humidity to thrive, thus accounting for the fact that only those individuals with plots near and within the premontane humid forest are able to grow it extensively. Second, while paja toquilla is valuable, it is also labor-intensive. Most landholders cannot afford to hire the additional labor required to expand their holdings. The distance of the forested areas from the settlements adds to labor costs.

The comuna itself places few constraints on landholders who want to cut down trees and grow more paja toquilla. Landholders enjoy secure rights to their land because they have been allocated plots by the comuna. No comuna rules exist to protect forested land from being cleared. Although the central government has recently banned commercial timber cutting and the hunting of deer in the forest, locals disregard the law since the government has only one forest guard for approximately eight comunas. Again, only the distance to the forest and the lack of capital to pay for additional labor constrain a rapid expansion of paja toquilla plantations. The cabildo is, in fact, ready to allocate another 5 ha to any of the forest landholders if they so desire.

The practices associated with the cultivation of paja toquilla thus help to explain the patchy condition of the forest in its southern parts. The forest's distance from the closest settlements (El Suspiro and La Ponga) encourages farmers to establish plantations in the part of the forest closest to their homes. The shortage of labor prevents these plots from being very large.

The final user group we consider is comprised of the two individuals who hold land in the forest but who make their livelihoods by cutting wood rather than growing paja toquilla. The woodcutters selectively cut the trees on their own plots within the forest; the vast majority of the wood they sell, however, comes from the trees they cut in the comuna reserve. Because the trees in this area are almost free of cost—besides the costs of traveling to the reserve and extracting the timber—the woodcutters choose to deplete this land first before they harvest from their own plots. Cutting from the communal plot also allows the trees on their land to "fatten" and thus become more valuable. The full-time woodcutters realize that they will be forced in the future to cut on their own plots to maintain their incomes. Demonstrating his belief that most of the valuable wood from the comuna reserve and individual plots will be removed relatively soon, one of the full-time woodcutters is "making connections" with members of another comuna in the hopes of either purchasing trees from its landholders or getting access to land to continue his occupation of cutting and selling trees. The other woodcutter is "networking" with larger commercial timber companies to the north of the comuna, hoping to ensconce himself as the middleman between them and furniture makers located in coastal towns.

Incentives of User Groups and the Lack of Institutional Demand

The management of Loma Alta's watershed could provide substantial benefits to comuna members. A management institution offers the possibility of sustainable product flows, which would provide a more secure long-term supply of timber and other forest products to individuals. The institution could help protect the integrity of the comuna's borders, thus ensuring that outsiders would not exploit comuna resources. And the institution would allow comuna members to continue to benefit from two critical public goods provided by the Loma Alta Forest: climate maintenance and watershed services (such as fog interception, the prevention of erosion, groundwater storage, and water purification).

Along with these benefits, however, the creation of institutions to protect a natural resource entails considerable costs. It is costly to reach agreement between the members of a community about what rules should regulate forest use. It is costly to structure monitoring efforts that ensure these rules are not broken. And it is costly to resolve the disputes that will arise when rules are broken.

The physical characteristics of a forest also affect the costs of organizing a management institution. The fact that the Loma Alta Forest is relatively distant from the four major settlements makes any monitoring effort by comuna members more difficult than if they lived adjacent to its borders. Additionally, members of other comunas can enter the Loma Alta Forest easily—the forest is not protected by natural or artificial barriers—increasing the likelihood of invasion and requiring more monitoring activities.

To cover these significant costs, the users of the forest must perceive significant benefits from forestry management to desire and to contribute to the creation of institutions to regulate the forest's use. While users of the Loma Alta Forest value the forest for certain products, it appears that members of these groups do not perceive the benefits of a managed forest to be greater than its costs.

Individuals who hunt game in the forest and those who purchase wood to build homes have little incentive to create an institution to regulate the forest's use. The small number of game hunters do not depend on the forest for any significant portion of their livelihood. While they would benefit from a well-managed forest, since it would likely contain more game, the hunters' stake in wildlife is relatively peripheral to their other daily activities.

Similarly, those individuals who use the forest's wood for constructing their homes have little incentive to shoulder the costs of forest management. While it is true that comuna members need wood to construct their homes and that they would likely have to pay higher prices for wood in the future if all of Loma Alta's trees were felled, individuals reap the benefit of inexpensive wood in the present. Wood from the open-access comuna reserve is there for the taking; wood from the plots of private landholders is still available. Even if the forest were completely denuded, Loma Alta's residents believe that other comunas could meet their timber needs. Given the benefit that most members enjoy from the current lack of timber restrictions, most would not favor—nor be willing to support—an institution that might restrict forest use.

Thus, both game hunters and wood purchasers use the forest intermittently, have available substitutes for the forest products they value, and do not depend on the forest for their livelihoods. These two user groups share a pattern of incentives that mitigates their desire to contribute time, effort, or money to manage the forest.

Paja farmers, timber cutters, and outsiders, in contrast, use the forest intensively, perceive fewer available alternatives, and depend on the forest and its products for a significant portion of their incomes. Paja farmers claim that if they could secure more labor or if the paths from their settlements to the forest were made easier to travel they would cut down more trees to plant more paja, their most valuable crop. Like the paja growers, the profitability of the woodcutters' activities depends on a consumptive use of the forest in the present. The woodcutters are already removing timber at a rate that presses them to plan for the day when the forest can no longer provide them timber to sell.

Neither paja growers nor woodcutters have an interest in institutional arrangements that restrict their use of the forest. Paja growers know that forest trees and paja plantations cannot coexist within the same plot; any limitations on the expansion of paja plantings would constrain their ability to increase their income. Woodcutters know that their use of the forest is nonsustainable. Their preference is to cut trees without restriction while trees still exist to cut. While their own plots within the forest may boast relative health, this may be an artifact of their ability to use the comuna reserve rather than a demonstration of any commitment to sustainable harvesting techniques. As long as the comuna reserve contains trees, woodcutters have the incentive to cut from that area first. When the reserve is completely denuded, it is likely that they will cut extensively on their own plots or in other comunas.

The outsiders who use the forest also favor the absence of forest regulations. The cattle-raising family has benefitted greatly from the fact that part of the comuna's forest remains open-access and unmonitored and from the lack of local institutions regarding forest use. In the absence of such institutions, the family has seized hundreds of acres. Like the paja farmers and the woodcutters, the outsiders' type of forest use—turning it into pasture—also threatens the forest's survival in the long term.

Significantly, only a few of the users are aware of the public goods provided by the forest; even fewer value these environmental services highly. Generally, comuna members have little knowledge of how the forest protects their watershed or affects their climate. While local nongovernmental organizations are trying to convince residents of various comunas in this region of the direct link between deforestation and the increasingly dry climate, paja growers and woodcutters do not mention these environmental concerns in discussions about their activities. Consequently, individuals value the forest for consumptive uses. And given the local economy and the rate of forest depletion, these consumptive uses appear unsustainable.

Conclusion

This study of the Loma Alta Forest highlights several issues regarding institutions, forests, and user groups important to policymakers concerned with Ecuador, as well as for scholars and practitioners interested in more general issues relating to the conservation of forests. The Loma Alta Forest shows deforestation rates that, if held constant, would result in total loss of trees on the remaining 950 ha in the next 25 years. On average over the past 20 years, 10 ha per year have been converted to paja toquilla and 30 ha per year to pasture. Maintaining the Loma Alta Forest is crucial to the entire community: loss of the multilayered forest will reduce water input to the groundwater resources of the Loma Alta watershed. With less forest cover, the vegetative surface area for intercepting moisture from the air is reduced, local evaporation is increased, and less water percolates down to aquifers. Both rainfall data and local memory confirm that Loma Alta's prolonged drought parallels the rate of deforestation, causing scientists, some officials, and locals to think the phenomena are closely related.

Despite the importance of the forest to the entire comuna, this study has shown that conceptualizing Loma Alta as a single entity, or viewing the forest as one resource, may not be fruitful methods by which to diagnose the causes of Loma Alta's deforestation. By viewing a forest as a resource that provides a number of different commodities and by examining the different groups who use these commodities, we provided an explanation for the lack of institutions regulating the Loma Alta Forest. While the comuna possesses most of the institutional assets that would favor the development of institutions, it has not yet created any rules regarding forest use. We found that those members with the biggest economic stake in the forest have no reason to limit their exploitative practices, and thus little demand exists for forest regulation at the local level. This lack of forestry institutions has led to an outcome whereby the Loma Alta Forest, while having some areas of relatively good secondary growth, is in danger of being more severely degraded in the near future.

Although no forest institutions exist in Loma Alta, we found that rules have had a direct impact on the forest's condition. The comuna's property-rights institutions, for example, provided a partial explanation for the pattern of forest use and current forest condition. As predicted by most property-rights theorists, the comuna reserve—that part of the forest without individual landholders—is the most seriously degraded (Demsetz, 1967; Libecap, 1989; North, 1990). Landholders, nonlandholders, and even noncomuna members choose to cut trees in the reserve first when they seek timber. Those plots with individual landholders, on the other hand, contain areas with less forest exploitation.

The Loma Alta case also demonstrates that strong individual property rights alone do not guarantee a forest's health. Landholders in Loma Alta possess incentives that do not favor the forest's long-term sustainability. Paja toquilla farmers would choose to expand their holdings of paja—which generates a certain and relatively long-term stream of income—over preserving the forest. Similarly, woodcutters earn income only with the removal of trees; even though their livelihood depends on some minimum population of trees, their short time horizons favor the complete removal of the trees before they consider a shift to other occupations.

To prevent continued deforestation in the Loma Alta area, policymakers must address the incentives that drive the behaviors of those users most crucial to the forest's existence—the farmers of paja toquilla, the woodcutters, and the outside invaders. Only when these actors consider alternative, less destructive activities to be of greater value than their present, more destructive practices will the forest's exploitation be limited. Part of the task confronting those interested in the long-term survival of the forest is to link comuna members' perceptions of the forest with its provision of public goods. If the forest's effects on the watershed and weather were more widely understood, locals may be more willing to support an institution that manages the forest's use.

Even if most comuna members highly valued the forest's public goods, however, there still remains a collective-action problem in the supply of institutions: although everyone benefits from the forest, it is an individual's interest to free-ride on the contributions of others (Olson, 1965; Ostrom, 1990). Given that no individual or small group in Loma Alta appears desirous of bearing the costs of starting a management institution, there may exist a role for nongovernment or government organizations to cover such start-up expenses (Thomson, 1992).

While considerable challenges confront those who wish to limit or stop Loma Alta's deforestation, the comuna possesses significant advantages over other rural areas. First, the population of the Loma Alta comuna is roughly stable. Approximately half of the young adults are leaving the area to pursue better employment opportunities in coastal urban areas.

The lack of population growth means that the pressure for farm land and timber may not increase rapidly in the near future. Second, the institutional assets of Loma Alta, discussed in the second part of this chapter, will be valuable to any attempt to construct a local solution to deforestation, despite the fact that the comuna presently has no institutions to regulate the use of their forest (Smale and Ruttan, 1995). The comuna's power to allocate property could be at the center of a policy that attempts to reserve land for watershed protection. The comuna's long history of member participation in committee building could facilitate the construction of monitoring and sanctioning devices as well as assist their staffing by comuna members. Finally, the comuna's experience with intragroup compromise will be critical to discussions that attempt to balance the goals of the comuna as a whole with members who stand to lose benefits if the comuna limits the use of its forest.

Acknowledgments

The authors would like to thank Carmen Bonifaz de Elao, professor of botany at the University of Guayaquil, for her help in directing, collecting, and analyzing forest-plot data; the support of Earthwatch and its volunteers for their hard work in carrying out this study; and Joby Jerrells, Miriam Lewis, Claude Nathan, Elinor Ostrom, George Varughese, and Rich Wolford for their constructive comments on this chapter.

Notes

1. We experienced this firsthand. We needed a small shelter to be built in the Loma Alta Forest to sample the flora of our random plots. When no one volunteered their forest land for the structure, an absent member's plot was chosen.

2. Two nonmembers, however, do hold title to private plots as a result of pre-1936 purchases. National and local governments respect the rights of those landowners whose purchases were completed before the enactment of the 1936 Law of the Comunas.

3. *Protective* refers to the forest's role in protecting the watershed.

4. Because the Loma Alta Forest is not above 2,000 meters, it cannot be defined as a typical cloud forest, although fog forests share much of the same characteristics (see Parker and Carr, 1992). See also the work of Dodson and Gentry (1991) on the forest resources of western Ecuador.

5. Such fog forests are of intense interest for those concerned with the conservation of biological diversity, since they boast endemic species and the conditions favorable for future speciation (Parker and Carr, 1992). Because the Chongon Colonche range is adjacent to, but geographically separated from, the Andes, its evolutionary pathways are isolated sufficiently to give rise to new subspecies and species. Conservationists are currently working on strategies to maintain evolutionary processes in these areas.

6. Comuna members are not clear about the borders of the area, and we were unable to survey the entire area. Further, the comuna does not possess a map of the reserve. Hence, the boundaries shown in figure 6.2 are our best estimate given discussions with comuna members but lacking the ground-truthing that we plan to undertake in the next phase of our research in the area.

7. The intensification of agriculture is not the result of population increases since the number of comuna members has remained fairly constant over the last two decades.

8. Coffee was formerly the most valuable crop in the region before drought and disease destroyed most plants in the area.

9. Because of the short duration of this pilot study, we sampled the areas of the forest considered the healthiest by comuna residents, forestry officials, and nongovernmental organization officials.

10. We sampled the plant communities in the fog forest to determine what biological influences both past and present uses of forest have had and to establish a baseline for monitoring the forest in the future. In this chapter, we focus on the condition of woody vegetation: trees, saplings, and seedlings in the forest. For this study, trees were defined as having a DBH $\geq$ 10 cm; samplings $\geq$ 2.5 cm but < 10 cm; and seedlings < 2.5 cm or a height of less than 1 m.

11. One extreme outlier, a *Ficus obtusifolia,* was omitted from the mean and standard deviation because of the difficulty in obtaining an accurate measurement (that is, discriminating between above-ground root system and trunk). The recorded DBH (200 cm) is nearly twice that of the next largest tree. The range of the sample is 100 cm.

12. For example, in a recent study of a Ugandan forest characterized as open access, over 50 percent of the plots had evidence of charcoal making, timber harvesting, or commercial firewood cutting (Becker, Banana, and Gombya-Ssembajjwe, 1995).

13. Three additional invaders have used land within the Loma Alta comuna, but each affects plots of less than 1 ha each.

14. The comuna allocated most of these areas to individual landholders in the 1960s and 1970s. This coincides both with the increasing dryness of lower comuna land and with demand for paja toquilla, which needs humidity to thrive.

References

Ascher, William. 1995. *Communities and Sustainable Forestry in Developing Countries*. San Francisco: ICS Press.

Becker, C. Dustin, Abwoli Banana, and William Gombya-Ssembajjwe. 1995. "Early Detection of Tropical Forest Degradation: An IFRI Pilot Study in Uganda." *Environmental Conservation* 22(1) (Spring): 31–38.

Blockhus, Jill M., Mark Dillenback, Jeffrey A. Sayer, and Per Wegee. 1992. *Conserving Biological Diversity in Managed Tropical Forests*. Gland, Switz.: International Union for the Conservation of Nature.

Bromley, Daniel, David Feeny, Margaret McKean, Pauline Peters, Jere Gilles, Ronald Oakerson, C. Ford Runge, and James Thomson, eds. 1992. *Making the Commons Work: Theory, Practice, and Policy*. San Francisco: ICS Press.

Demsetz, Harold. 1967. "Toward a Theory of Property Rights." *American Economic Review* 57 (May): 347–59.

Dodson, C. H., and Alwyn H. Gentry. 1991. "Biological Extinction in Western Ecuador." *Annals of the Missouri Botanical Garden* 78: 273–95.

Fundacion Natura. 1992. "Proyecto Cordillera Chongon-Colonche: Inventario y Diagnostico Fisico." Unpublished report. Fundacion Natura, Guayaquil.

Gibson, Clark, and Stuart Marks. 1995. "Transforming Rural Hunters into Conservationists: An Assessment of Community-Based Wildlife Management Programs in Africa." *World Development* 23: 941–57.

Hecht, Susanna, and Alexander Cockburn. 1990. *The Fate of the Forest: Developers, Destroyers, and Defenders of the Amazon*. New York: Harper.

Libecap, Gary D. 1989. *Contracting for Property Rights*. New York: Cambridge University Press.

Marks, Stuart. 1984. *The Imperial Lion: Human Dimensions of Wildlife Management in Central Africa*. Boulder, CO: Westview Press.

McCay, Bonnie J., and James M. Acheson. 1987. *The Question of the Commons: The Culture and Ecology of Communal Resources*. Tucson: University of Arizona Press.

North, Douglass C. 1990. *Institutions, Institutional Change, and Economic Performance*. New York: Cambridge University Press.

Olson, Mancur. 1965. *The Logic of Collective Action*. Cambridge: Cambridge University Press.

Ostrom, Elinor. 1990. *Governing the Commons: The Evolution of Institutions for Collective Action*. New York: Cambridge University Press.

Parker, Theodore A., and John L. Carr. 1992. "Status of Forest Remnants in the Cordillera de la Costa and Adjacent Areas of Southwestern Ecuador." RAP Working Papers 2, Conservation International, Washington, DC.

Poffenberger, Mark, ed. 1990. *Keepers of the Forest: Land Management Alternatives in Southeast Asia.* West Hartford, CT: Kumarian Press.

Richards, P. W. 1975. *The Tropical Rain Forest.* Cambridge: Cambridge University Press.

Schlager, Edella, and Elinor Ostrom. 1993. "Property Rights Regimes and Coastal Fisheries: An Empirical Analysis." In *The Political Economy of Customs and Culture: Informal Solutions to the Commons Problem,* ed. Terry L. Anderson and Randy T. Simmons, 13–41. Lanham, MD: Rowman & Littlefield.

Smale, Melinda, and Vernon W. Ruttan. 1995. "Cultural Endowments, Institutional Renovation and Technical Innovation: The *Groupements Naame* of Yatenga, Burkina Faso." Unpublished paper, University of Minnesota.

Thomson, James. 1992. *A Framework for Analyzing Institutional Incentives in Community Forestry.* Community Forestry Note No. 10. Rome: FAO.

United Nations Food and Agriculture Organization (UNFAO). 1990. *The Community's Toolbox: The Idea, Methods and Tools for Participatory Assessment, Monitoring and Evaluation in Community Forestry.* Rome: FAO.

7

Indigenous Forest Management in the Bolivian Amazon: Lessons from the Yuracaré People

C. Dustin Becker and Rosario León

Introduction

Societies have been making choices about their relationships with forests for many centuries. As reviewed by Perlin (1991), the dominant choice for the last 5,000 years across Asia and Europe, and more recently in the Americas, has been to cut down trees, use them for fuel and building materials, and replace them with crops or urban centers. In contrast, numerous neotropical cultures have evolved societies with informal norms that sustain rather than destroy forest ecosystems (Chernela, 1989; Posey, 1992). Such ecologically oriented cultures are rapidly disappearing. Mutualistic relationships between forests and people in the tropics are changing as activities in the forest are modified by incentives structured by market forces, government forest policies, and changes in the values of indigenous peoples. This study explores the changing relationship between the Yuracaré people and the forest communities they sustain and use along the Chapare River in northern Bolivia. It finds that while several external threats affect the condition of the Yuracaré's forests, a significant amount of the forest and its biological diversity still benefit from traditional rules for human activities in the forest.

Historical and Ecological Setting

In the early 1990s, national policy in Bolivia shifted from ignoring the rights of indigenous people in the Amazon to taking them into consideration. The policy change came about in response to internal land-tenure conflicts, political organization by indigenous groups, and lobbying by human rights and conservation groups (Paz et al., 1995). Years of conflict between families indigenous to the Amazon and settlers from more populated regions of Bolivia led to protests by native groups. In 1990, indigenous Amazonians of Bolivia staged a march "for their territories and dignity" (Paz et al., 1995). This political unrest, combined with the decentralization policies of international donor agencies, prompted the Bolivian government to overturn the Law of Colonization. Promulgated in 1966, this law declared the lands of the Amazon to be uninhabited and open for colonization. With the negation of the old law, indigenous groups are now recognized and are currently being given legal authority over their traditional territories. The government plays a supervisory role by evaluating petitions for land tenure from indigenous groups in the Amazon.

As a prerequisite for acquiring title to the lands and waters they have used for the past 400 years, the Yuracaré are required to create a management plan for the stewardship of the natural resources within their traditional boundaries (CERES, 1997). This requirement implies that the Yuracaré lack forest management, a supposition that has never been questioned or explored. It has also invited external assistance and influence in defining and creating a modern management plan. In this chapter, we assess whether the Yuracaré truly lack a system of forest management and what sort of socioeconomic forces are likely to influence them as they try to forge a forest-management plan that will be acceptable to government decision makers.

The Yuracaré people have one of the last remaining forests in Bolivia that is clearly under indigenous control. Approximately 400 Yuracaré families live in the northeastern part of the Department of Cochabamba. They currently claim about 250,000 hectares of the Chapare River watershed as their territory. In 1994, the Yuracaré began a collaboration with the Forests, Trees, and People Programme (FTPP) of the United Nations Food and Agriculture Organization (FAO) based at the Centro de Estudios de la Realidad Economica y Social (CERES). The goal of this collaboration was to make an official forest-management plan that would be acceptable to the Bolivian government.

CERES conducted an International Forestry Resources and Institutions (IFRI) study to provide an initial understanding of the relationship between people and forests in the Yuracaré culture (CERES, 1997). Three settlements along the Chapare River—Misiones, Trinidadcito, and Santa Anita—and their associated forests were studied by CERES (figure 7.1). The settlements are located in three "life zones" (Holdridge, 1967), all of which may be broadly classified as lowland tropical moist forest. Misiones is positioned in the life zone referred to by Holdridge (1967) as "wet tropical forest." Here rainfall ranges from 2,200 to 4,400 millimeters per annum, and temperature ranges from 17 to 24 degrees Centigrade. Trinidadcito is in "moist tropical forest" and receives between 1,900 and 2,800 mm of rain each year, and temperatures remain relatively constant, 22 to 24°C. Santa Anita receives 1,250 to 1,450 mm of rain each year and thus supports a forest that is transitional between dry and moist (Holdridge, 1967).

The Chapare River has as much influence on vegetation communities as rainfall. Alluvial soils have been deposited at all sites studied by CERES, and moisture and erosion have established a riparian forest community that is fairly homogeneous in species composition along the entire river. The most common tree genera are palms, *Astrocaryum* and *Scheelea,* and fruiting hardwoods, *Guarea, Inga, Rhipidocladum, Theobroma* (wild cocoa), *Virola,* and *Hura* (CERES, 1997). In disturbed areas, early successional trees genera, *Inga* and *Cecropia,* dominate the vegetation community.

Forest inventories indicate that the potential for timber extraction in Yuracaré territory was as high as 49 cubic meters per hectare (Rojas, 1996). Bolivia's Department of Forestry (DIDF) set quotas on timber extraction in the Chapare River region based on these estimates and has encouraged the Yuracaré to organize forest associations to meet these quotas. In response to this marketing incentive, the Yuracaré organized their own forest associations and privatized the most valuable timber.

Themes, Definitions, Null Hypotheses, and Corollaries

In this chapter we use data from the CERES-IFRI study to explore whether the Yuracaré have a tradition of forest management. To address

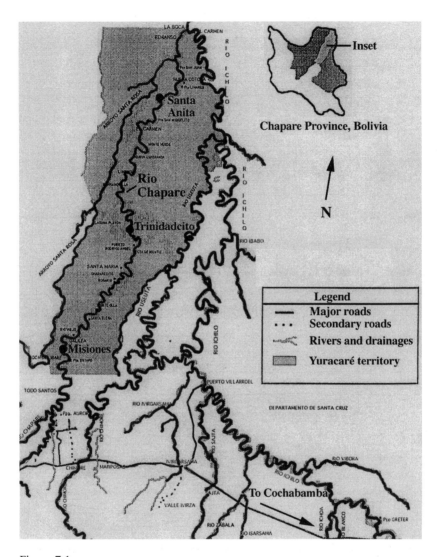

Figure 7.1
Location of Yuracaré territory, settlements, and forests on the Rio Chapare, Bolivia

this question it is also necessary to examine what other important factors may be influencing the condition of the forest. Consequently, in addition to studying Yuracaré institutions that might affect the stands of trees that make up the forest, we also investigate the effect of moisture gradient, population pressure, and distance from market. Because the CERES-IFRI study includes data from both the natural and social sciences, we examine factors from both of these traditions. We use social science data to determine whether there are social norms in place that are directly aimed at forest stewardship. We use data from forest stand inventories to look for physical evidence of forest management by the Yuracaré. The intensity of forest use varies along the river. By comparing the riparian forest at three sites that vary greatly in their distances to market and intensity of use (population pressure), we can begin to assess the effects of the ingress of the Bolivian market economy on the forest and the Yuracaré's relationships with it.

Consistent with theory developed in the IFRI research program (see the appendix to this volume), we define *forest institutions* as rules or social norms applied to forest goods and services. A *rule* is considered to be a social regularity with deontic content (implication of "must" or "must not") that is observable, interpretable, or explainable by a local person (Ostrom, 1992). The Institutional Analysis and Development (IAD) framework, on which IFRI is founded, considers actions taken at several levels of social organization: operational, collective choice, and constitutional (Ostrom, 1990). In this case, the operational level refers to actions of individuals that affect the state of the forest (harvesting, transplanting, pruning, culling, and so on). Collective choice applies to actions of individuals that affect the operational level (prescribing, invoking, monitoring, enforcing, and so on). Finally, actions at the constitutional level affect collective choice by determining who prescribes, invokes, monitors, or enforces rules.

We pose the following null hypothesis: Yuracaré people living on the Chapare River of Bolivia lack forest-management institutions. If this is true, and drawing from the design principles of Ostrom (1990), it follows that the Yuracaré will *lack* constitutional, collective choice, or operational activities that

- Prevent destruction of important forest resources,
- Encourage activities that conserve or restore forest resources, and
- Clearly define boundaries (Ostrom, 1990) and access to forest resources.

These are all typical forest-management activities or norms that sustain forests (Aplet et al., 1993). Sustainable forests may result from either constraining use or from reforestation, and long-term resource management may contain elements of both strategies. Boundaries may be organized at many levels, such as individual (rights to specific trees), family (areas managed by a group of relatives), and regional (use defined by membership in an indigenous group).

Borrowing from cultural anthropology, the null hypothesis also predicts that the Yuracaré will lack language pertaining to forest management, especially regarding aspects of sustainable use such as long-term planning and constraints on individual use. More specifically, the null hypothesis predicts that the Yuracaré will demonstrate

- Little knowledge about forest resources in language or traditions,
- No awareness of resource depletion, or actions to remedy it, and
- No conceptual awareness of the role of individual constraint in sustaining a natural-resource common.

Again these predictions relate directly to the potential of any society to sustain a biological resource base while they use it. The managers must have basic knowledge about the distribution and abundance of the resource to be managed and knowledge of how that resource reproduces and grows. They must also be able to detect depletion and to modify use in such a way that the resource can recover or hover around some equilibrium population size that sustains use over long periods of time.

Forest condition along the Chapare River should reflect ecology and human utilization as influenced by population density, market demand, and the social institutions that control use of forest products (figure 7.2). Because forests are relatively old systems, a long time horizon must be considered. Current forest biomass and diversity may reflect decisions and actions made decades (even centuries) in the past as well as those made recently. To evaluate the past and present social impacts on for-

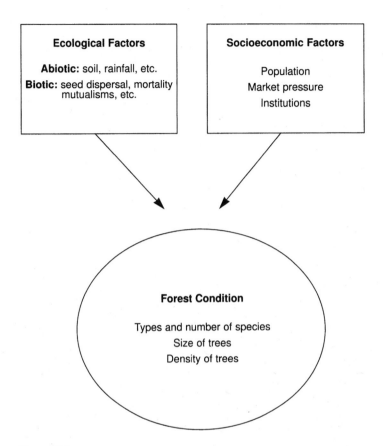

Figure 7.2
Social and ecological factors that influence forest condition

ests, we use the paradigms of forest ecology. Thus, forest condition is defined by measurable physical and biological aspects of a plant community dominated by trees and other woody plants. Measurements include but are not limited to central tendencies for biomass, basal area, species diversity, density of woody stems, canopy cover, as well as spatial distributions of disturbance, ground cover, and particular plant species (IFRI, 1997).

Biomass of trees and species diversity vary in response to population pressure, market demand, and according to forest-management rules. However, moisture gradients and stochastic patterns of seed dispersal and herbivory are equally viable explanatory factors for variation in forest structure and composition (Spurr and Barnes, 1980). How does one determine when forest condition reflects societal (institutional) outcome rather than ecological pattern? This sort of challenge can be solved with a research design that varies institutions over a similar ecology or a design that partitions ecology and sociology. In this case, we use the rainfall gradient (moisture) to distinguish the structuring forces of ecology and human use.

Moisture-Gradient Hypothesis

We assume continuity in regional climate over the life of the forest (last 200 to 300 years), so the rainfall gradient documented for the Chapare River should affect basal area, species abundance, and distribution. As mentioned above, Misiones receives more precipitation per year, on average, than Trinidadcito, and Santa Anita receives the least. Trees should thus have largest diameters at breast height (DBH)[1] in Misiones, where rainfall is plentiful, while progressively smaller values should be found in Trinidadcito and Santa Anita for within-species comparisons. Species diversity should also be highest at Misiones and progressively lower as moisture constraints come into effect downstream.

Population-Density Hypothesis

Of the approximately 1,800 human inhabitants along the Chapare River, populations of about 600 are permanent at Misiones and Santa Anita. In contrast, Trinidadcito has no permanent settlement, and thus population pressure has been historically low there relative to the other two settlements. If density-dependent effects of resource utilization influence the forest, one would expect a pattern of low-high-low for measures of density (trees per hectare) and basal area of trees in Misiones, Trinidadcito, and Santa Anita, respectively.

Market-Demand Hypothesis

Opposite to the effects of moisture gradient, commercial timber species should show an increase in density and basal area with distance from Cochabamba because market pressure declines with distance from a major trading center. The largest trees will be harvested where the cost to get them to market (distance) is the least. Commercial tree species should show a pattern of low-medium-high for basal area and density in Misiones, Trinidadcito, and Santa Anita, respectively.

In addition to being influenced by moisture gradient, population density, and market forces, forested areas around each settlement have been partitioned into "communal" and "family-managed" units. Following the logic popularized as the "tragedy of the commons" (Hardin, 1968), communal forests would be expected to have lower densities and basal areas, while family-managed forests should be better conserved and stocked. Table 7.1 summarizes the ecological and socioeconomic variables hypothesized to influence forest condition and reveals that each alternative hypothesis has its own mutually exclusive outcome.

After a brief description of method we compare the average density, DBH, and basal area of tree species used for commercial timber, domestic timber, and fruits and medicines and see which factors best explain their current distribution and abundance.

Methodological Details and a Reference Forest

Institutional analysis and forest-stand inventories—standard methods of the IFRI research program (see the appendix to this volume)—were

Table 7.1
Predicted variation in forest condition at Rio Chapare sites in response to social and ecological factors

Forest Structuring Factor	Relative Mean Values for Stem Density and Basal Area		
	Misiones	Trinidadcito	Santa Anita
Moisture gradient	Large	Medium	Small
Human population pressure	Small	Large	Small
Market pressure	Small	Medium	Large
Tenure	Family plots > Community forest plots at all three sites		

Note: *Stem density* refers to the number of trees in a given area. *Basal area* is a measure of the woody biomass in a given area (usually per hectare).

completed for five forest sites (CERES, 1997). Information about social norms and institutions was obtained during visits with Yuracaré families and at larger community gatherings, using participatory rural appraisal (PRA) activities and informal discussion. Data were entered on standardized IFRI forms. Forest-plot data (IFRI, 1997) were aggregated by communal and family forest areas at Misiones and Santa Anita, but such tenure differences were not in place at Trinidadcito (not a permanent settlement).

Within each forest stand, trees were sampled in circular plots with a 10-meter radius. Plots were systematically placed at 100-m intervals along 1-kilometer transects perpendicular to the river and exclusively in mature riparian forest. Areas cultivated with annuals and monocultures of perennials like bananas were purposefully excluded from the forest-sampling effort. Transects were positioned in stands of trees that have remained under use by the Yuracaré over the past few centuries. Sample sizes were stratified according to the size of the forest remaining near each settlement. Because biological diversity typically increases with area sampled, we compared richness within 1-hectare areas, unless reported otherwise. We use species and family richness (number of tree species and families per hectare) as a measure of biological diversity in the different forest sites.

To examine the human impact on the riparian forests of the Chapare River and place it into context for the western Amazon region, we compared our data with those describing the riparian forests of Manu, Peru, a large protected area (Gentry and Terbourgh, 1990; Foster, 1990). In this case, Manu's stands serve as a control of sorts or as a reference forest because these forests have been protected from timber exploitation and have not been used intensively by indigenous families for at least four decades. They are in the same major Amazon watershed and share similar ecologies. We predict that the riparian forests used by the Yuracaré will differ from Manu as follows:

• Basal area of trees will be consistently lower at all three settlement areas than at Manu (due to Luman use).

• Diameters at breast height (DBH) will be consistently smaller along the Yuracaré than at Manu (due to exploitation of large-diameter timber species).

• Tree diversity on the Chapare River will be substantially lower due to human use.

Results

Forest-Management Institutions Created by the Yuracaré

The Yuracaré have clearly defined boundaries, systems of monitoring resource condition, and rules that directly pertain to forest resources. Yuracaré institutions control access and use of the forest at multiple levels: clan (extended family), *corregimiento*,[2] and territory. Tribal territory and clan areas are largely predicated on providing families within clans with sufficient game meat and other natural resources. Clans are the core of the Yuracaré social system, consisting of an extended family made up of 10 to 20 nuclear families (husband, wife, and children). Clans have organized themselves into 11 corregimientos along the river. Within each corregimiento, *kuklete* ("family forest gardens") are created, cared for, and monitored like private property. Families state that they "own their work" but not the land. While territorial tenure is important for the Yuracaré people-forest relationship, permanent private landholdings are not, because families strategically move within their corregimiento and within the entire territory, creating forest gardens and obtaining forest resources that vary in availability spatially and temporally. Thus, each family has a stake in organizing to maintain control over the whole watershed, and their institutions reflect this landscape-level concern for sustaining their resource base (table 7.2).

Using a consensus approach, representatives from each clan elect a *Cacique Mayor Yuracaré*[3] to lead them. Likewise, each corregimiento has several representatives that participate in a tribal council (*Consejo Indigena Yuracaré*). This council uses a system of one-person-one-vote and majority rule to make major decisions and plans that concern the Yuracaré society as a whole. Since territorial control is a major concern of the Yuracaré, council meetings tend to focus on political conflicts with other indigenous groups and on interactions with external agencies (such as government, church, and nongovernmental organizations) in relation to land tenure.

Table 7.2
Framework of Yuracaré forest institutions

Social Level	Operational	Collective Choice	Constitutional
Individuals, families, clans	Planting fruit trees Culling for fruiting trees Protecting fruiting trees Harvesting timber trees	Allocating land to families Choosing where to develop family tree gardens Leaving the communal forest	Defining clan membership Familial decision making Selecting clan leaders
Corregimiento (subregion)	Monitoring resource use Sanctioning abusers	Allocating land and forest to different clans Deciding on ownership of commercial tree species	Being members of a forest association Clan leaders comparing use within their areas
Territory (watershed)	Monitoring commercial timber Sanctioning abusers	Families interacting with clan leaders and *cacique* (Yuracaré chief) to resolve tree tenure conflicts Meetings held when needed	Clan leaders comparing use within their areas with input from forest association and government

Most forest-management activities (operational) and decisions (collective choice) are made at the family level within clans. Families have an informal normative system for monitoring resources and their use within their corregimiento. The approach is completely decentralized but replicated within the entire territory. Information about resource distribution—such as the location of timber species, excellent hunting areas, trees in fruit, and areas that are good for cultivation, is well known in each corregimiento. Families in each corregimiento do informal inventories of resources by walking and canoeing throughout their region and discussing the spatial distribution of resources. Over the centuries this knowledge has been culturally systematized and used to classify soils, to design a system of forest agriculture, and, more recently, to exploit commercial timber (Paz et al., 1995).

Exploitation of commercial timber created conflict and challenged Yuracaré institutions because at first certain individuals accrued more benefits than others. In response, clans devised a system of tree ownership to distribute this wealth more equitably. In 1991, forest associations were formed in each corregimiento to organize timber exploitation and to interact with government forestry departments and timber buyers.

Both constitutional and collective-choice levels of organization are represented by social norms that prescribe actions pertaining to forest management in Yuracaré culture. For example, a frequently mentioned norm was that "All Yuracaré must care for the forest." When asked why, the typical response was "So the animals will come." Rights and obligations are thus created and enforced by the Yuracaré as a group. Operationally, "caring for the forest" includes protecting fruiting trees and transplanting and selectively encouraging fruiting trees to increase densities of game. Yuracaré also have game-management rules including selective harvesting of the males and no-hunting seasons. When rules are broken, sanctioning is traditionally accomplished via social reprimand and ostracism, but with the growth of commercial timbering, these mild social sanctions have been inadequate at times. For example, a man harvested and sold trees belonging to another family, and this required conflict resolution between two clans. The norm breaker was required to split his income from the sale of the trees with the original owner.

Well before the organization of forest associations, Yuracaré language and traditional norms included explicit prescriptions for sustainable forest management. At the collective-choice level, the Yuracaré prescribe "use of forest trees and animals without depletion." This prescription is operationalized through a "mobile multiple-use" relationship with the forest and through the creation of fruit-tree gardens. Rather than using any one forest area intensively, families spread out the impact of timber harvesting, agriculture, hunting, and gathering in time and space. Movements include complex local and regional patterns, a full description of which is beyond the scope of this chapter. However, seasonal variation in resource use and collection of resources over a large area can be interpreted to prevent depletion of patchy resources in any one area. The Yuracaré practice long-term biodiverse perennial agriculture in small forest patches. Areas with productive soils, *ti jukule,* are first planted with yucca, then bananas, and then fruit trees (such as mango, chocolate, orange, coffee, grapefruit, palms, and native fruit trees). The forest-tree garden is used for 25 to 35 years, eventually via succession becoming mature rain forest, dominated by domestic and wild fruiting species. The Yuracaré promote growth of the wild fruiting species by culling nearby seedlings of nonfruit trees.

Yuracaré institutions are highly responsive to external incentives. In 1992, the Yuracaré decided to organize forest associations in each corregimiento to coordinate with external government forest agencies and timber marketing associations. Two laws—the Forest Law and the Law of INRA (National Institute of Agrarian Reform)—have given the Yuracaré exclusive rights to forest exploitation within their territory but under constraints and directives imposed by government forestry agencies. Issues of concern to the forest associations include equitable allocation of resources, controls on harvesting methods (such as lobbying to relax rules against use of chainsaws), and resolution of conflicts among themselves. To reduce conflicts over valuable timber, the Yuracaré forest associations privatized mahogany and Spanish cedar in community forest areas. Since these forests are already rather equally distributed among family areas via the corregimiento system, the private goods within them were relatively easy to distribute.

Yuracaré Language and Forest Management

The Yuracaré have many sayings that explicitly relate to sustaining a diverse tropical-forest ecology. Their language is replete with statements about the Yuracaré's role in a food chain based on fruiting trees. The following phrases are translations illustrating a linguistic familiarity with the ecological concept that forest fruit feeds game animals and that game animals provide food for the Yuracaré:

1. "To be human one must eat meat." → *carne*, "flesh"
2. "When the ambaibo (Cecropia) fruits, the animals get fat!"
3. "Yuracaré must care for the forest."

In addition to stating that people should "use forest trees and animals without depletion," the Yuracaré have the saying *"Cuivalimatu tëpshë dulashtututi nomajsha"* (One should plan for the future). Such language illustrates a familiarity with conservation principles that underlie sustainable use.

When asked to name natural resources, wild forest fruits and animals had a higher proportion of indigenous names than timber and agricultural species (figure 7.3). This result is not surprising given that Yuracaré culture was totally dependent on forest resources prior to colonial influence. What is more important is that the Yuracaré named 52 fruiting tree species that they actively monitor, protect, and promote. The Yuracaré have spatial concepts of their forest resources as evidenced by maps they made during participatory rural appraisal (PRA) exercises (figure 7.4). While their geographic information system (GIS) may lack precision (such as not being to scale), it accurately depicts locations of forest resources, forest-cover classification, water resources, and use patterns.

Variation in Forest Condition

As shown in table 7.3, forest conditions at the three sites on the Chapare River differ substantially. Despite having higher rainfall that would favor large mean basal area and good regeneration, Misiones had the lowest values for both of these important indicators of forest health. Forests in Misiones, where timber exploitation was heaviest, had a basal area of only 28 m^2/ha, while stands in Santa Anita averaged 38 m^2/ha.

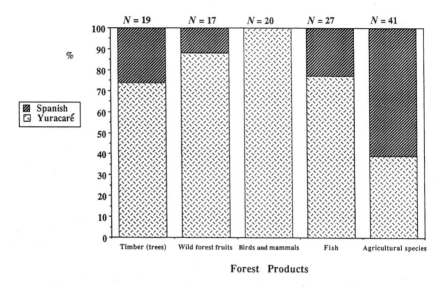

Figure 7.3
Proportion of different natural resources with Yuracaré and Spanish names
Note: Wild forest fruits, birds, and mammals have higher proportions of indigenous names than timber trees, fish, and agricultural resources.

In general, the mean basal area of trees along the Chapare River increased with distance from market (table 7.3). Basal area was not consistently lower on family plots, nor was it lower than the mean basal area at the Manu River protected area. Average basal area was largest for family plots at Santa Anita, a result best fitting predictions for distance from market. Consistent with an outcome based on timber exploitation, diameter at breast height was larger at Chapare River sites than at Manu River sites (table 7.3). Tree diameters were significantly smaller at Misiones than at Trinidadcito and Santa Anita (ANOVA, $df = 4$, $p = 0.02$).

Mean density of trees varied as a consequence of use along the Chapare River and was only half the value for the Manu River. While the Chapare River forests had from 310 to 366 trees per ha, Manu had 650 trees per ha. When the distribution of size classes are compared for the two river systems (figure 7.5), a pattern consistent with market incentives and traditional forest management may be interpreted. Trees with diameters of 26 to 40 cm were clearly less abundant in the Chapare River sample than

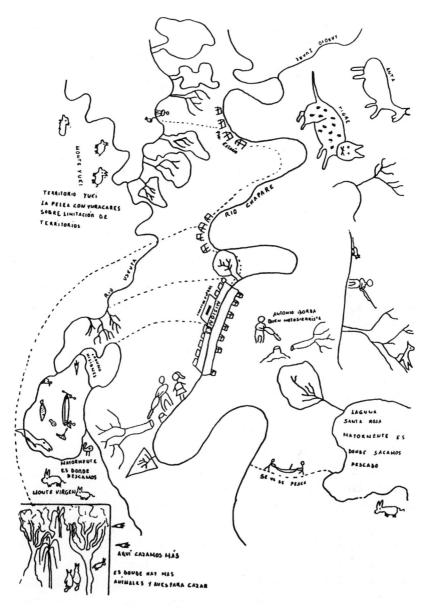

Figure 7.4
People-forest relationship map made by Yuracaré living in Misiones
Note: Illustration emphasizes a utilitarian relationship with the forest ecosystem,
including use of chainsaws for timber harvesting (Paz et al., 1995).

Table 7.3
Tree basal area, density, and diversity at Rio Chapare sites

| | Misiones | | Trinidadcito | Santa Anita | | Rio Manu |
	Familiar	Comunal		Familiar	Comunal	Cocha Cashu[a]
Annual precipitation (mm)	2,280–4,400		1,900–2,800	1,250–1,450		2,028
Basal area (m²/ha)	27	29	33	40.3	36	37
Trees/ha	319	333	363	310	366	650
Mean DBH ± se	24.7 ± 1.2	24.4 ± .7	27.1 ± .45	27.8 ± 2	26.4 ± 1.3	22.6[b]
Tree diversity						
Tree species/ha			45–60[c] (for all sites)			155–283
Tree families/ha			23–29 (for all sites)			45
Tree families > 10 ha			34 (for all sites)			N/A
% palms[d]	2	8	17.5	1.8	1.5	15.3
Local land use	Market and family		Mobile family	Family and ranching		Tourism

Note: Comparison with "pristine" Rio Manu forest trees ≥ 10 cm DBH. All sites are on alluvial deposits.
a. "Pristine" upper Amazon alluvial floodplain forest (Gentry and Terbourgh, 1990).
b. Calculated by taking midpoint values of size classes from Gentry and Terbourgh (1990, table 27.1).
c. Lower value may result from difficulty of identification at species level, although family count is also lower.
d. *Astrocaryum, Iriartea, Scheelea.*

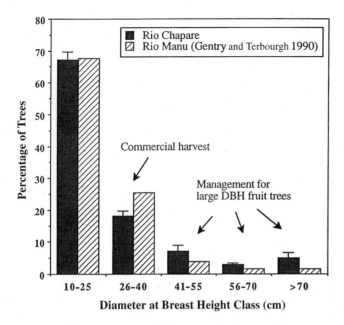

Figure 7.5
Comparison of size classes (diameter per breast height) of trees in forest stands on the Rio Chapare and on Rio Manu

at Manu, probably a consequence of timber harvest. Large-diameter fruit trees, however, were more abundant in the Chapare River sample than in the Manu sample. There were also more trees per ha in the communal forests than in family forests (not significant at the plot level), quite unlike a "tragedy of the commons" scenario.

Tree species diversity along the Chapare River was low relative to Manu (table 7.3). While botanists found as many as 283 different species in 1-ha samples at Manu, the maximum value at Chapare River sites was 60 species. Comparisons made at the family level (where identification skill is less likely to bias results) also suggest that forests along the Manu are more diverse than those associated with the Chapare River. Trees in 1 ha at Manu represented 45 families, while only 34 families were found in the Chapare River study in an area of more than 12 ha.

DBH of trees used indigenously averaged 10 cm larger than commercial species, suggesting that market pressure has lowered the biomass of timber species, while species used for fruit, local building material, and medicines have been conserved (table 7.4). Nine of the 10 most abundant tree species in the Chapare River samples were fruiting species used by birds and mammals that are traditional foods of the Yuracaré. Still, several noncommercial species such as *Amendrillo* and *Crespito* had very low regeneration values, and *Yesquero* had no evidence of regeneration (table 7.4). Two timber species of traditional importance, Gabun (*Virola peruviana*) and Guayabochi (*Calycophyllum sproceanum*), show little regeneration in the Misiones forest samples, suggesting that they may be overexploited there. Seedlings and saplings of two medicinal trees, Paraquina (*Ephedranthus amazonicus*) and Gabetillo (*Sloanea rufa*), were also nearly absent in the Misiones plots.

Of the 34 tree species with economic importance to the Yuracaré, nine exhibited changes in density and DBH that would be expected along a gradient of soil moisture (table 7.5). When biomass profiles (DBH and stem density) are compared (table 7.5), commercial timber species were more likely to be depleted than traditional timber species. None of the traditional timber species showed the "low-medium-high" profile consistent with market pressure. Given the lack of market profiles in the more abundant traditional use species, it is possible that "depleted" species in this category are rare species. Eleven of 28 species with traditional uses (table 7.5, panels B and C) showed reductions in density and diameter consistent with population pressure. Three species had similar average values at all sites, and only two species had profiles that did not fit any predicted pattern.

Commercial timber species had smaller diameters than fruit trees despite their low regeneration statistics (figure 7.6). No fruiting species showed indications of depletion, while traditional timber species did. This suggests that sustainable stewardship is directed to those species that directly contribute to the Yuracaré food chain, while timber species are not managed in a sustainable fashion.

Discussion and Conclusions

The distribution and abundance of tree species along the Chapare River is complex but reflects moisture gradients, population pressure, and market

Table 7.4
Estimates of trees ha^{-1} of commercial and noncommercial timber species on the Rio Chapare, Bolivia (genera in parentheses)

	Trees (per hectare)	Mean Percentage	Average Diameter at Breast Height	Saplings (per hectare)
Commercial Species				
Trompillo (*Guarea*)	13	3.5	19	122
Gabun (*Virola*)	12	3.2	32	48
Verdolago (*Terminalia*)	6.1	2.3	33	11.8
Laurel (*Ocotea*)	1.9	<1	22.4	11.8
Palo Maria (*Calophyllum*)	.74	<1	36.5	.9
Cedro (*Cedrela*)	.14	<1	15.6	1.6
Mara (*Swietenia*)	0	—	—	0
Total ~34				
Mean = 26.4				
Noncommercial Species				
Jorori (*Swartzia*)	6.7		25.4	2.7
Guayabochi (*Calycophyllum*)	2.2		30.2	3.6
Yesquero (*Cariniana*)	0.4		66.0	0
Uropi (*Claricia*)	2.6		23.3	13.6
Almendrillo (*Dipterex*)	0.2		99.0	1.8
Ochoo (*Hura*)	9.2		45.3	9
Negrillo (*Nectandra*)	4.7		20.4	30
Cedrillo (*Spondias*)	4.1		39.5	4
Cafesillo (*Margaritaria*)	7.3		29.8	21
Coloradrillo (*Brysonima*)	8.6		19.6	71
Crespito (*Stryphnodendron*)	1.5		24.0	1
Sangre de Toro (*Virola*)	3.5		24.7	3.6
Coquino (*Pouteria*)	4.7		25	10.9
Total ~56				
Mean = 36.3				

Note: Data from the five IFRI forests were pooled because there were no statistical differences in tree or sapling densities by site. Estimates are derived from 386 plots totalling 12.12 ha. Sapling estimates are based on a 1.1-ha aggregate of 386 plots, each covering 28.3 m^2. Sampling was stratified by forest size, so these data are biased toward Trinidadcito, where 57 percent of plots were completed.

Table 7.5
Biomass profiles

Species	Misiones Density C	Misiones Density F	Misiones DBH C	Misiones DBH F	Trinidadcito Density	Trinidadcito DBH	Santa Anita Density C	Santa Anita Density F	Santa Anita DBH C	Santa Anita DBH F	Result: Major Effect
A. Commercial timber species											
Cedrela sp.	0	0	0	0	1	16	0	0	0	0	Depletion
Dipterex odorata	1	0	150	0	1	48	0	0	0	0	Depletion
Guarea sp.	14	12	17	22	15	20	7	9	23	13	Moisture
Hura crepitans	5	7	42	45	12	51	7	13	30	57	Market
Swietenia macrophylla	0	0	0	0	0	0	0	0	0	0	Depletion
Terminalia amazonica	9	12	36	22	5	57	8	5	50	114	Market
B. Traditional timber species											
Annona sp.	5	9	55	26	6	57	2	0	27	0	Moisture
Brysonima indorum	19	9	18	16	6	19	13	5	32	117	Family use
Calycophylum sp.	3	1	30	17	3	30	1	0	49	0	Family use
Cariniana estrellensis	0	1	0	61	1	71	0	0	0	0	Rare or depleted
Ceiba pentandra	1	0	92	0	0	0	0	0	0	0	Rare or depleted
Claricia racemosa	2	2	18	25	4	32	1	2	21	20	Population
Margaritaria nobilis	8	13	25	23	8	27	7	4	12	21	Similar
Nectandra sp.	7	4	26	21	5	22	7	16	2	17	Family use
Ocotea sp.	1	3	19	33	3	21	0	1	0	10	Moisture
Pouteria bilocularis	1	0	14	0	7	29	1	1	36	19	Family use
Pouteria sp.	3	1	26	15	0	0	0	10	0	10	Moisture

Species											Major effect	
Stryphnodendron sp.	1	1	13	22	1	23	1	1	21	31	Similar	———
Virola peruviana	15	17	25	17	13	22	0	0	0	0	Moisture	–'–
Virola sebifera	5	1	26	20	0	0	14	17	19	33	Family use	–'–

C. Tree Species with traditional use for fruits (F) and medicines (M)

Species											Major effect	
Astrocaryum chonta (F)[a]	18	6	16	17	42	17	36	33	15	21	Population	–'–
Brosimum lactescens (F)	2	2	17	11	9	17	2	0	20	0	Population	–'–
Cordia nodosa (M)[a]	0	0	0	0	0	0	2	1	26	14	Market	–'–
Ficus insipida (M)	0	0	0	0	2	21	0	0	0	0	Population	–'–
Ficus sp. (M)	10	17	48	59	5	60	4	1	86	120	Moisture	'–'
Inga sp. (F)	44	37	20	19	17	18	14	12	18	21	Moisture	'–'
Leonia glycicarpa (F)	5	3	13	14	10	15	0	1	0	27	Other	–'–
Machura sp. (F)	6	5	12	15	1	15	0	0	0	0	Moisture	'–'
Scheelea princeps (F)	5	1			23	9	13				Population	–'–
Sloanea rufa (M)[a]	1	0	20	0	1	19	6	4	20	17	Market	–'–
Spondias mombin (F)	4	2	43	21	5	45	4	4	51	37	Similar	–'–
Theobroma cacao (F)	1	2	19	13	5	16	26	15	15	15	Other	---
Theobroma speciosum (M)	5	2	14	19	11	18	5	1	12	12	Population	–'–
Triplaris americana (M)	5	4	13	15	2	16	0	0	0	0	Moisture	'–'

Density (stems per hectare) and mean DBH of tree species in different use categories in communal (C) and family (F) forest plots in three settlements along the Rio Chapare, Bolivia. The last column presents a verbal and pictorial representation of biomass or importance (a combination of density and DBH data). For example, the species *Hura crepitans* shows an increasing value in density and mean DBH across a row (by settlement) and is thus represented by the pictograph (–'–). This profile fits our predictions for market exploitation. In cases where density and DBH appear to decline in family forests, the term *family use* is placed in the major effect column. *Similar* indicates that no hypothetical cause can be identified.

a. Also used for building materials.

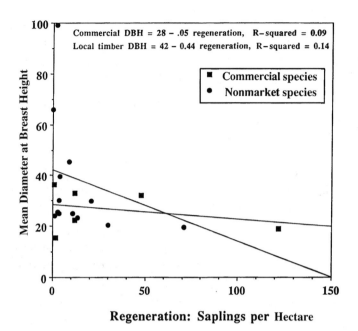

Figure 7.6
Linear regression of mean diameter at breast height on regeneration
Note: Commercial species appear to be less conserved than noncommercial species. Species with poor regeneration value and no market pressure appear to be left to attain very large diameters before they are harvested, while commercial species lack this trend (see table 7.5 for details of different species).

demand in predictable ways. It was possible to detect single factors that determine the abundance of trees along a river gradient and to determine when social constraints outweighed ecological ones. It is extremely clear that timber marketing is changing forest structure along the Chapare River.

The results of this study refute the idea that the Yuracaré lack forest-management institutions. In addition to cultural norms that prevent destruction of important forest resources, encourage activities that conserve or restore forest resources, and define boundaries and access to forest resources, the Yuracaré make and modify rules for forest use (Ostrom, 1990). They also monitor physical condition and use of forest resources, sanction abuse of forest resources, and resolve conflicts over those forest resources. Clearly, the Yuracaré have constitutional, collective action, and operational activities that explicitly pertain to forest management, and their forest-management system existed well before external government forestry agencies began to demand forest-management plans. The Yuracaré also have language and traditions that would be considered hallmarks of sustainable forest management: long-term planning and constraints on individual use.

All along the Chapare River the Yuracaré have reduced tree density and diversity, but their selection for large fruiting trees has increased basal area and biomass. Because fruit abundance is positively correlated with basal area and diameter at breast height (Leighton and Leighton, 1982), Yuracaré forest management should enhance resources for wildlife. Essentially, their traditional forest management is a mutualism with fruiting trees and game animals. The Yuracaré increase the reproductive success of fruiting trees, which increases the density of game, which has potential to increase Yuracaré survival via food availability.

Although their long history of self-organization has helped them respond efficiently to recent incentives for timber exploitation, their tradition of conservation of fruit trees has not been extended to timber management. They are capable of integrating with government forestry agencies to negotiate harvest quotas and constraints on harvest technology, but they show little inclination to conserve or restore timber species. Perhaps it is just too early to judge, but our data suggest that several species have been extirpated in the entire region and that extirpation of traditional species is now occurring around Misiones where market demand and alternatives to traditional cultural patterns are greatest. Traditional interest in fruiting trees and dependence on forest resources is also changing as families depend less on local resources and enter the market economy.

Paz (1991) suggested that timber extraction would have negative consequences for the Yuracaré because the loss of timber species might cause a collapse in traditional resources used for subsistence. At first glance, a collapse in forest resources seems unlikely because the Yuracaré appear to be maintaining fruiting trees and other trees that sustain their valued food chain and traditional needs. The timber species are surplus goods in the Yuracaré way of life. The Yuracaré "mobile multiple use" system also buffers against depletion of common-pool resources in the forest.

On closer inspection, however, our results support Paz's conjecture. The parasitic relationship of timber exploitation is beginning to erode the Yuracaré's mutualistic relationship. Biomass and diversity are being lost as market and institutional incentives favor the unsustainable harvest of timber species. Privatization of forest resources may also promote a more sedentary life, which would put more pressure on local resources. That this is happening is supported by the differences in forest condition at Trinidadcito and the two permanent settlements. Presumably, Paz was thinking about the breakdown of ecological links between timber and fruiting species that sustain wildlife habitat and other important mutualisms (pollination, dens for game animals, and so on). The ecological links between timber and fruiting species that might sustain wildlife are unknown.

Misiones, where population and market pressures are greatest, showed the greatest declines in basal area and abundance of individual tree species relative to other forests studied on the Chapare River. If the Yuracaré turn from hunting and tending forest gardens to embracing market economies and timber-management incentives, the fruiting forest would play less of a role in their culture and language, and a timber-producing forest would become more important. Forests along the entire Chapare River could become as degraded as those around Misiones.

While Yuracaré folk ecology refutes our null hypothesis, this does not imply that indigenous people have all the knowledge and institutional capacity required to manage forest resources. In this case, conservation end points were only convergent for certain tree species (fruiting trees). Ecological cognition prompts the Yuracaré to be more risk adverse toward substitutions for fruiting species, while they are less adverse to liquidation of timber species. Timber harvesting has tested Yuracaré tradition and deserves more study from a socioanthropological context. For example, it is not clear how the different forest associations have resolved conflicts over privatization of timber species or if any of the clans will institute a system of sustained-yield harvesting for timber trees. Given their traditional biases, they should monitor regeneration and focus timber exploitation on nonfruiting species that regenerate well. Will they protect their traditional fruiting trees or buy into the market economy?

Comparisons of colonial and indigenous settlements in the Amazon have consistently shown that forest degradation is typically greater under stewardship of colonial farmers than under the care of forest dwellers with a long history of interacting with forest resources (Rudel, 1993; Chernala, 1989; Atran et al., 1998). Encouragingly, some colonists in the Petén of Guatemala have adopted ecological values and actions of forest-adapted Itzaj Mayans rather than dissuading the indigenous people from forest sustaining practices (Atran et al., 1998). Atran optimistically interprets this finding as potential for indigenous knowledge to influence policy and planning at regional and national levels, and advocates that coevolved relationships between indigenous people and forest species should be considered more carefully by policymakers.

The Yuracaré have not fully evaluated the possible negative consequences of forest degradation, nor have they evaluated the bargaining power that might result from conserving their trees. The increase in basal area or biomass resulting from traditional Yuracaré forest management can be viewed as a positive contribution to carbon storage, which is a global commons benefit. Ecological cognition and values of the Yuracaré seem to be somewhat limited to the fruit trees that form the base of their food pyramid, yet their system has sustained people, wildlife, and a diverse forest for many centuries. Timber extraction has had negative impacts on both carbon storage (basal area) and biodiversity. A forest policy based on commercial timber poses uncertainties for the sustainable aspects of a 400-year-old relationship between people and forests along the Chapare River. Regional planners in Bolivia and external agents promoting timber harvesting need to monitor the environmental impacts they are having on the mutualistic strategies inherent in the traditional Yuracaré forest management.

Acknowledgments

We thank Julie England and Robin Humphrey for assistance with data compilation. Clark Gibson, Fabrice Edouard Lehoucq, and Elinor Ostrom kindly critiqued early drafts of the chapter. Funding for fieldwork and the opportunity for the authors to collaborate was provided by the FAO, the Ford Foundation, CIPEC, CERES, and the Workshop in Political Theory and Policy Analysis of Indiana University. We recognize the following IFRI researchers who made this synthesis possible: José Antonio Arrueta, Nelson Castellón, Daniel Chávez Orosco, Freddy Cruz, Antonio Guzmán Suárez, Fernando Miranda, Ignacio Nuñez Ichu, Juan Carlos Parada Galindo, Antonio Patiño Salazar, Benancio Rodríguez Bolivar, Miguel Rodríguez Chávez, Patricia Uberhuaga, and Galia Vargas.

Notes

1. This standard measurement of tree diameters is taken at 1.4 meters and is used to calculate basal area $= \pi * (DBH/2)^2$.

2. *Corregimiento* is related to the noun *corregidor,* which refers to a Spanish magistrate. In this case, the term applies to a spatially defined unit of governance organized by one or more Yuracaré clans.

3. *Cacique* means "political leader." It is also the name of colorful, loud, social birds in the neotropics from which feathers are used to decorate leaders.

References

Aplet, G. H., N. Johnson, J. T. Olson, and V. A. Sample, eds. 1993. *Defining Sustainable Forestry.* Covelo, CA: Island Press.

Atran, S., D. Medin, R. Ross, J. Coley, E. Lynch, E. U. Ek', P. Kockelman, and V. Vapnarsky. 1998. "Folkecology and Commons Management in Maya Lowlands." Manuscript.

Centro de Estudios de la Realidad Economica y Social (CERES). 1997. *Manejo comunal del bosque y gestión del territorio Yuracaré.* Cochabamba, Bolivia: FTPP-FAO.

Chernela, J. M. 1989. "Managing Rivers of Hunger: The Tukano of Brazil." *Advances in Economic Botany* 7: 238–48.

Foster, R. B. 1990. "The Floristic Composition of the Río Manu Floodplain Forest." In *Four Neotropical Rainforests,* ed. A. H. Gentry. New Haven: Yale University Press.

Gentry, A. H., and J. Terbourgh. 1990. "Composition and Dynamics of the Cocha Cashu 'Mature' Floodplain Forest." In *Four Neotropical Rainforests,* ed. A. W. Gentry, 542–64. New Haven: Yale University Press.

Hardin, G. 1968. "The Tragedy of the Commons." *Science* 162: 1,243–48.

Holdridge, L. R. 1967. *Life Zone Ecology.* San Jose, Costa Rica: Tropical Resources Center.

International Forestry Resources and Institutions (IFRI). 1997. *IFRI Field Manual. Version 8.0.* Bloomington: Indiana University, Workshop in Political Theory and Policy Analysis.

Leighton, M., and D. R. Leighton. 1982. "The Relationship of Size and Feeding Aggregate to Size of Food Patch: Howler Monkey *Alouatta palliata* Feeding in *Trichilia cipo* Trees on Barro Colorado Island." *Biotropica* 14: 81–90.

Ostrom, E. 1990. *Governing the Commons: The Evolution of Institutions for Collective Action.* New York: Cambridge University Press.

————. 1992. *Crafting Institutions for Self-Governing Irrigation Systems.* San Francisco: Institute for Contemporary Studies Press.

Paz, S. 1991. *Relaciones interétnicas en las nacientes del río Mamoré.* Cochabamba: UMSS.

Paz, S., M. Chiquueno, J. Cutamurajay, and C. Prado. 1995. *Estudio Comparativo: Arboles y Alimentos en Comunidades Indigenas del Oriente Boliviano.* Cochabamba: CERES/ILDIS.

Perlin, John. 1991. *A Forest Journey: The Role of Wood in the Development of Civilization.* Cambridge: Harvard University Press.

Posey, D. A. 1992. "Traditional Knowledge, Conservation and 'The Rainforest Harvest'." In *Sustainable Harvest and Marketing of Rain Forest Products,* ed. M. Plotkin and L. Famolare. Washington, DC: Conservation International, Island Press.

Rojas, C. 1996. Personal communication.

Rudel, T. K. 1993. *Tropical Deforestation: Small Farmers and Land Clearing in the Ecuadorian Amazon. Methods and Cases in Conservation Science.* New York: Columbia University Press.

Spurr, S. H., and B. V. Barnes. 1980. *Forest Ecology.* New York: Wiley.

8

Population and Forest Dynamics in the Hills of Nepal: Institutional Remedies by Rural Communities

George Varughese

Introduction

Projections of massive declines in Himalayan forest cover and dire predictions for the future of forests in Nepal initiated worldwide concern in the 1970s (Eckholm, 1975, 1976; World Bank, 1978). Initially, the source of the problem was seen as domestic fuelwood use compounded by rapid population growth. Then expansion of agriculture, commercial logging, and tourism were blamed. However, the actual rates of deforestation, as well as its causes and consequences, remain very much in question. Studies indicate that while there is degradation from overharvesting in the hills, the total loss of forest cover has been relatively small (for example, Ives and Messerli, 1989). Others argue that losses have even been reversed in both forest area (HMG, 1988; Bajracharya, 1983; Metz, 1990; Gilmour and Nurse, 1991) and tree density (Messerschmidt, 1986; Gilmour and Fisher, 1992). Still others contend that while forest area is not decreasing in the hills, the quality of existing forests is suspect (Chakraborty et al., 1997; Subedi, 1997).

This debate notwithstanding, the future remains insecure and disturbing for Nepal's rural majority who depend on forests. Even though the claims of dire environmental crisis might have been exaggerated, rising population, migration, increased industrial and commercial activity, and developmental pressures continue to place heavy demands on the forest resource base. In a country where over 80 percent of the population depends entirely on agricultural and forest products for food, fodder, and fuel, forested lands always face the risk of being used at an unsustainable rate. Consequently, the issue of how to best govern forest resources in Nepal remains of critical concern to policymakers.

Population change lies at the heart of this debate, as it does for resource-management and -development policy globally. While for many population growth is accepted as a primary or intermediary cause of re-source degradation (Ehrlich and Ehrlich, 1991; Brown, Wolf, and Starke, 1987; Bilsborrow and DeLargy, 1991), for others an increasing popula-tion is a stimulus to economic development and innovative resource-management practices (Boserup, 1965, 1981; Simon, 1981, 1983, 1990; Binswanger and Pingali, 1989). In general, it has been difficult to find agreement on what the relationship is between population growth and natural-resource condition.

This study examines the relationship between the governance of forest resources and population in the middle hills of Nepal. Specifically, it in-vestigates the significance of local institutions in forest resource manage-ment to gain a better understanding of how such institutions shape the actions of individuals at the community level. By focusing on local institu-tions, this study becomes less concerned with what or who is the agent of environmental degradation than with what has helped forest users to cope with environmental and population change. Indeed, for the 18 loca-tions in this study, the findings indicate that change in forest conditions is not significantly associated with population growth. Rather, change in forest conditions is found to be strongly associated with local forms of collective action. This implies that policymakers' preoccupation with population growth as a primary determinant of resource degradation may be ill-advised. Instead, the facilitation of institutional growth and innova-tion at the local level may be more relevant to the robustness of the natural-resource base.

The first section of this chapter provides a general overview of the on-going debate about the relationship between population growth and the environment. This overview provides the backdrop for a review in the second section of research that addresses forest resources in Nepal. The third section provides a description of the research setting and the approach used to conduct the study. The fourth section introduces the variables used for the study and reports the findings for the 18 locations. The fifth section provides a closer look at a set of six cases selected to understand differences in physical outcomes across the 18 locations. The chapter concludes with a discussion of some of the key factors that help explain differences between communities that have coped with population and resource change.

Population and the Environment

A great deal of research has focused on the relationship between population change and the environment, and the debate continues. Since Malthus, scholars have argued forcefully that population growth is the primary cause of environmental degradation (Abernathy, 1993; Brown, Wolf, and Starke, 1987; Ehrlich and Ehrlich, 1991; Myers, 1991; Wilson, 1992). While demographers in this tradition have shown that population growth has some negative consequences, others have shown that population growth can also lead to technological advances and innovative uses of natural resources (Simon, 1983, 1990; Boserup, 1965, 1981; Binswanger and Pingali, 1989). Increasingly, research addressing the relationship between population change and the environment demonstrates that their linkages are complex and yet to be understood fully (Bilsborrow and DeLargy, 1991; Cruz et al., 1992; Jolly, 1994; Netting, 1993; Shivakoti et al., 1997). While it is clear that demographic change does influence resource use, population growth is but one variable of a larger set of important variables whose numerous interactions affect the natural-resource base.

Part of the difficulty in understanding the linkages between population change and the environment is that, methodologically, much of the extant research examines agents of environmental change at a high level of aggregation. By resorting to a macro perspective, most of these studies have handicapped their ability to exploit micro-level research to understand the complex workings of population and environment linkages (Arizpe, Stone, and Major, 1994). Scholars of microinstitutional solutions to commons problems have long argued that local communities can craft durable institutional arrangements that enable them to successfully manage local natural resources, even when confronted with political, economic, and demographic pressures (Acheson, 1989; Feeny et al., 1990; Ostrom, 1990). These scholars recognize, however, that successful local solutions are more difficult to achieve where (1) demographic change is rapid, (2) a local community is not dependent on the resource in question, (3) substantial heterogeneities of interest exist, (4) little local autonomy exists to make and enforce rules, and (5) the resource system itself is very large (see, for example, Ostrom, 1998b). Thus, studying how local communities cope with different kinds of population pressures is a major topic of theoretical and policy interest.

In more focused research on factors that mediate environment-population interactions in the Kumaon Himalaya of India, Agrawal and Yadama (1997) have argued that by studying micro relationships at the community level it is possible to gain an understanding of how variables such as population, economic growth, and forest area get aggregated at a macrostructural level. Their study of 275 rural communities finds that local institutions play a critical role in mediating demographic and socioeconomic influences.

This study explicitly recognizes that factors such as population change can influence resource use in a variety of ways. But rather than be determinative of human behavior, the study investigates how resource users might craft institutional arrangements to cope with demographic and environmental forces.

Research on Nepal

The growth of population and its supposed effect on the Nepali Middle Hills has been the subject of several studies. The earliest and most influential was conducted by Eckholm (1975, 1976), who drew attention to population growth in the Nepali hills and rather tenuously linked it to "denuded hillsides" and "deteriorating environments" where "the pace of destruction is reaching unignorable proportions" (1975, 764–65). Subsequently, it was shown that this connection between an increase in population and catastrophe in the hills was simplistic and misleading (Bajracharya, 1983; Ives, 1987; Ives and Messerli, 1989; Mahat, Griffin, and Shepherd, 1986a, 1986b).

In addition to rapid population growth, government policies of nationalization in the 1950s and 1960s have been identified by most researchers as one of the main causes of deforestation. Placing the ownership of forests with the national government disrupted preexisting and traditional practices of communal resource management. Since the government lacked sufficient human or economic resources to look after newly nationalized forests, what was once communally governed property became open to anyone to exploit. Traditional management practices that have endured and more recent innovative community forestry legislation, on the other hand, have been credited for enabling the forest conservation and regeneration that has taken place in the Middle Hills since the 1960s (Arnold and Campbell, 1986; Mahat, Griffin, and Shepherd, 1986a, 1986b, 1987a, 1987b; Messerschmidt, 1986; Griffin, 1988; Hobley, 1990; Exo, 1990; Gilmour and Fisher, 1992; Chhetri and Pandey, 1992; Dahal, 1994; Pradhan and Parks, 1995; Subedi, 1997).

Recent studies of Nepal's forest-management practices have directed attention toward the importance of institutional arrangements and social mechanisms. Some researchers have pointed to the role played by local institutional arrangements in sustainable resource use (e.g., Gronow and Shrestha, 1991; Gilmour and Fisher, 1992), but none have undertaken a study of institutional arrangements and their mediating effects on resource conditions. In a similar vein, studies have incorporated some descriptions of institutional arrangements within detailed descriptions of forest-user groups (Chhetri and Pandey, 1992; Dahal, 1994; Karki, Karki, and Karki, 1994; New ERA, 1996). While this work represents progress in Nepali forestry research, there is a paucity of social scientific research that brings an institutional approach to the study of local forms of community organization in forestry.

While the population in the Middle Hills continues to grow close to an annual rate of 2 percent at present, its effects on the surrounding patchwork of forest land are not so clear. One reason has been the absence of longitudinal data on forest condition and forest use. Few researchers have studied the same location over time. One notable exception is the study conducted by Jefferson Fox in a Nepali village in the Middle Hills in 1980 and 1990. Fox found that forest conditions were improved substantially, even though population density increased significantly over a period of ten years. Fox's finding had little to do with the dynamics of population parameters. Rather, changes in the authority of villagers to manage nearby forests, the construction of a road that reduced the costs of inputs needed to adapt traditional agricultural practices, and the provision of external help in the form of knowledge rather than financial aid appeared to be the most important factors for improved forest conditions (Fox, 1993). Clearly, population parameters alone did not drive these outcomes.

Another reason for the lacuna in research on forest condition and use in Nepal has been the lack of consistently collected cross-sectional data (Subedi, 1997). Frequently, the inherent weaknesses of a study done in a single time period can be overcome if a sufficient number of similar studies are done using the same research methodology and theoretical framework in a single time period. This study seeks to address this gap in knowledge by looking at local-level information on demographic and forest parameters across several locations in the Middle Hills visited in a single time period.

The Study Setting

The physiographic zone of the Middle Hills of Nepal provides the broad setting of this study. In the Middle Hills, the population is estimated at 8.4 million (45.5 percent) with a growth rate of 1.61 percent for 1981 to 1991 (Central Bureau of Statistics, 1995). (Nepal's total population was 18.5 million with an annual growth rate of about 2.08 percent for the same time period.) The population remains largely rural, with fewer than 10 percent of the total in towns and cities. Subsistence agriculture is still the main occupation, although villagers do not hesitate to supplement their livelihoods by entering the market economy whenever opportunities arise.

The rural population in the Middle Hills is mainly distributed in small villages or hamlets that are sometimes parts of larger, dispersed settlements. A common pattern of forest-land distribution in these hills is for small patches of forests to be scattered throughout larger areas of cultivated land. These are vital sources of fuelwood, fodder, and leaf litter for animal bedding and composting, especially in the winter months when agricultural residues are exhausted. In 1985 to 1986, forest land (of about 5.5 million hectares) accounted for a substantial proportion (38 percent) of the total land area (about 14.7 million ha) in the country. The Middle

Hills contained about 1.8 million ha (32.6 percent) of forest land in this time period (HMG, 1988).

The change in use of forest resources in the hills has not been ascertained with any accuracy. However, a recent study of over 3,300 households in Nepal found that 93.7 percent of rural households collected firewood, and 86.8 percent used firewood as cooking fuel. Of all the households collecting firewood, 25.3 percent collected from their own land, 12.5 percent collected from community forest land, 59.7 percent used government forest land, and 2.6 percent obtained firewood from other sources (Central Bureau of Statistics, 1996). Evidently, nonprivate forest lands continue to supply the majority of firewood for households in the hills, upwards of 70 percent. The figures for community and government forest-land usage are only useful in estimating nonprivate land use. Frequently, what is officially government land is actually communal by use. The figures also do not supply acreage of various lands used for forest products. It could well be that the community forests and private lands are less used because of management regimes in effect.

Community forestry in the Middle Hills is being implemented through the administrative structure of the Department of Forests, facilitated by various donor-aided programs. These range in size from bilateral projects covering one or two districts (such as the Nepal-Australia Community Forestry Project) or seven districts (the Nepal-UK Community Forestry Project) to the largest (the Community Forestry Development Project), which is providing technical assistance and financial support, by way of World Bank assistance, to 35 hill districts. The 18 sites included in this study are from districts in the Middle Hills, most of which have various sorts of community-based integrated-development program activities, including the community forestry program of the Nepali government.

A Study of Eighteen Cases in the Middle Hills of Nepal

To examine the roles of institutions and population in forest-resource change, this study employed a two-stage analysis. The first stage of analysis provides a broad understanding of trends in population changes and the association of these trends with (1) foresters' and villagers' perceptions of forest conditions (changes in tree density and in forest area) and

(2) evidence of local-level organization and cooperation in resource management in the set of 18 cases. The second stage of analysis focuses on six cases that help illustrate the patterns discerned in the initial analysis. The task is to identify and examine how the crafting and operation of institutional arrangements generate different outcomes.

The cases included in this study are shown in table 8.1 in the chronological order in which they were visited by the International Forestry Resources and Institutions (IFRI) research program team in Nepal. These cases comprise a larger set of IFRI studies conducted in various physiographic zones of Nepal since 1992. The data for these particular cases were obtained over a period of three years. Each case was studied by a five-member team comprised of natural science and social science researchers over a period of four weeks using IFRI research methods (see Ostrom, 1998a; see the appendix to this volume).

The 18 cases in this study represent locations within village development committees (VDCs) in the Middle Hills of Nepal and range from the easternmost district of Ilam in the Eastern Development Region to Gorkha and Tanahun districts in the Western Development Region (see figure 8.1). For the purposes of this study, the names of settlements are omitted, and instead, locations are identified using the names of the VDC within which the settlements and forests were studied. All but two of the studies (Manichaur and Sunkhani) conducted in the Western and Central Development Regions are part of a series commissioned by the Hills Leasehold Forestry and Forage Development Project of the government to monitor the effect of the project in those locations over time. As part of that monitoring plan, some of these locations have already been revisited since the first round of baseline studies; other locations are being revisited in the spring of 1998. The Manichaur and Sunkhani locations were studied as baseline assessments of forest-use patterns in the Shivapuri Integrated Watershed Development Project north of Kathmandu valley.

In the Eastern Development Region, the cases are part of a longitudinal series of IFRI studies, funded by the MacArthur Foundation, that examine forest resources and institutions in locations that have varying access to markets and roads and that are in areas of high and low intervention by government and donor agencies. Thus, the locations of study were

Table 8.1
Descriptive statistics for 18 sites

Site Location	Date of Visit	Population		Average Household Size	Forest Area (hectares)	Forest Stock Assessment[a]
		Individuals	Households			
Churiyamai VDC (Makwanpur)	March 1994	4,500	750	6.0	85	Average
Baramchi VDC (Sindhupalchowk)	May 1994	244	36	6.7	75	Below average
Riyale VDC (Kavre Palanchowk)	May 1994	644	92	7.0	29	Average
Bijulikot VDC (Ramechhap)	June 1994	980	145	6.7	53	Average
Thulo Sirubari VDC (Sindhupalchowk)	April 1995	843	105	8.0	16	Average
Doramba VDC (Ramechhap)	May 1995	139	26	5.3	107	Average
Agra VDC (Makwanpur)	June 1995	434	70	6.2	190	Average
Bhagwatisthan VDC (Kavre Palanchowk)	June 1995	471	70	6.7	108	Below average
Manichaur VDC (Kathmandu)	June 1996	1,550	242	6.4	115	Average
Sunkhani VDC (Nuwakot)	September 1996	1,065	144	7.4	290	Below average
Chhimkeshwari VDC (Tanahun)	December 1996	192	28	6.8	45	Average
Chhoprak VDC (Gorkha)	January 1997	781	106	7.4	25	Below average
Raniswara VDC (Gorkha)	February 1997	2,661	404	6.6	300	Average
Bandipur VDC (Tanahun)	February 1997	1,021	183	5.6	75	Above average
Barbote VDC (Ilam)	May 1997	1,467	260	5.6	145	Average
Shantipur VDC (Ilam)	May 1997	162	29	5.6	90	Average
Chunmang VDC (Dhankuta)	June 1997	922	152	6.1	225	Average
Bhedetar VDC (Dhankuta)	June 1997	477	82	5.8	125	Above average

Note: Names in parentheses are districts.
a. Assessed by a forester based on tree density and speciation during the period of study and cross-checked where possible with district forest officials.

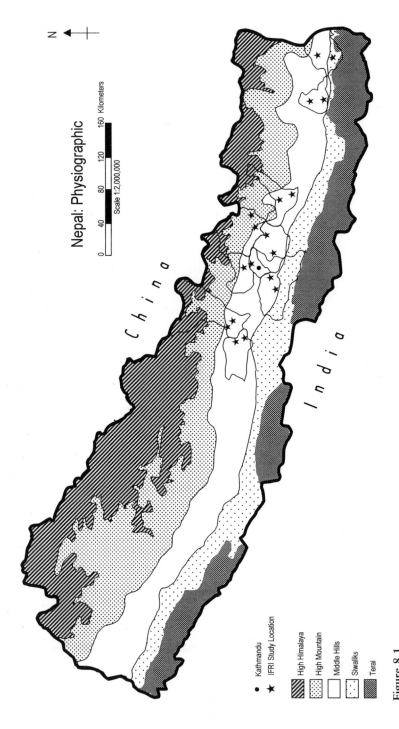

Figure 8.1
Eighteen locations in districts visited for International Forestry Resources and Institutions studies in Nepal

mainly determined on the basis of project or agency criteria. However, the data obtained show variation on the factors I examine in this study—the indicators of population growth and change in forest conditions and the degree of collectively organized activity by forest users.

The study initially uses descriptive indicators such as household and individual population, average household size, and forest area and stock condition to provide some idea of the locations visited (table 8.1). In particular, the indicator *forest stock* provides a subjective assessment of forest condition at the time of the study by the forest specialists on the research team with respect to speciation and abundance of vegetation. In most of the 18 cases, the professional assessments of the district forest officials in those study sites were also obtained to validate the research team's subjective assessment. This assessment also gives researchers an initial idea of the natural endowment that each group of users possesses. By itself, this assessment is not a good longitudinal indicator of forest condition, but when combined with some measure of change in forest condition (see table 8.2), one is able to obtain a general picture of resource-use patterns and management.

At the time of this study, forest data were still being compiled from revisits to several of these locations, and, therefore, the indicators used here for forest condition are limited to those based on assessments made by villagers and foresters. In other IFRI studies, more rigorous measures of vegetative stock are used in addition to measures based on assessments by villagers and foresters (see, for example, Becker, Banana, and Gombya-Ssembajjwe, 1995; Varughese, 1999).[1]

In the 18 locations studied, household and individual population, average household size, and forest area exhibited considerable variation (table 8.1). The number of individuals in a group of forest users varied from 139 to 4,500, and the number of households per group varied from 26 to 750. Across the sites studied, this gives a range of 5.3 to 8 individuals per household for average household size across the sites studied. The average household size across all 18 locations is 6.43 individuals per household. In comparison, a recent survey by the Central Bureau of Statistics (CBS) on Nepal living standards found the average household size to be 5.33 in this physiographic zone (CBS, 1996). The area of forest land used as a primary source of forest produce by villagers in these

Table 8.2
Preliminary comparisons of population growth with forest condition

Site Location	Population Growth Rate (percent)	Households per Hectare	Trend in Forest Condition[a]
Doramba (Ramechhap)	7.37	0.24	Improving
Churiyamai (Makwanpur)	5.42	8.82	Improving
Shantipur (Ilam)	5.22	0.32	Worsening
Bhedetar (Dhankuta)	5.14	0.66	Worsening
Raniswara (Gorkha)	4.71	1.35	Improving
Chunmang (Dhankuta)	4.13	0.68	Worsening
Baramchi (Sindhupalchowk)	4.00	0.48	Stable
Barbote (Ilam)	3.64	1.80	Stable
Bijulikot (Ramechhap)	3.39	2.74	Improving
Riyale (Kavre Palanchowk)	3.00	3.17	Stable
Sunkhani (Nuwakot)	2.68	0.50	Worsening
Bhagwatisthan (Kavre Palanchowk)	2.60	0.65	Worsening
Chhoprak (Gorkha)	2.55	4.24	Worsening
Manichaur (Kathmandu)	2.28	2.10	Improving
Thulo Sirubari (Sindhupalchowk)	2.11	6.56	Stable
Bandipur (Tanahun)	1.44	2.44	Improving
Agra (Makwanpur)	0.29	0.37	Worsening
Chhimkeshwari (Tanahun)	−1.33	0.62	Stable

a. Assessed by villagers based on local historical understanding and corroborated, in most instances, by district forest officials.

locations varied from 16 ha to 300 ha with an average across sites of 116.56 ha. The condition of most of these forests was found to be within the average range in this physiographic zone. Only two locations had above-average stocks, and three had below-average stocks. This assessment is made relative to typical forest stocks to be found in this zone as determined by the Department of Forests.

Table 8.2 provides comparisons of population growth rate, average households per hectare of forest area, and trend in forest condition. The population growth rate is obtained by taking the difference in households (from the time of the visit to five years prior) and averaging it over five years. The five-year rate is preferred here because the assessments of forest condition in this study are also based on a five-year period. The 10- and 20-year growth rates were also available but are used only to supplement the discussion. The trend in forest condition is a subjective assessment of forest condition derived from the historical perceptions of diverse local forest users and, in many instances, of local government forest officials, about the relative abundance of produce, disappearance of valuable species, and change in forest area: "worsening" indicates a clear depletion of species and reduction in forest area; "improving" indicates at least a perceptible increase in abundance of tree species and shrubs. The locations are arrayed from high to low rates of population growth in table 8.2.

Table 8.2 is more useful in understanding changes for each site and provides some interesting findings. In general, the population growth rates (averaged over five years) vary from a negative growth rate of -1.33 to well over 7 percent per annum with a range of 8.70 and a mean of 3.26 percent per year. For a 10-year period, the growth rates vary from 0.37 to 10 percent per annum with a range of 9.63 and a mean of 4.08 percent per year. It is important to note that these growth rates are well above the national average for this physiographic zone, calculated to be 1.61 in 1991 (CBS, 1995). The household-to-forest ratios in these locations also exhibit dramatic variation, from 0.24 to 8.82 households per hectare of forest area with an average of 2.10 households per hectare. These figures show that there can be considerable variation from place to place in demographic characteristics across a physiographic zone.

However, is this variation reflected in forest condition? Across the 18 locations, there are six forests in improving condition, five in stable condition, and seven in worsening condition. But if the growth rate is taken as a first demographic measure, the two highest rates (7.37 and 5.42) seen in Doramba and Churiyamai have a forest stock that is average and improving. The lowest rates (-1.33 and 0.29) seen in Chhimkeshwari and Agra have a forest stock that is average in condition but is stable (in Chhimkeshwari) or worsening (Agra). Furthermore, if the number of households per hectare of forest available is taken as a

Table 8.3
Association of population growth with forest condition

Forest Condition	Population Growth		Total
	Above Average	Below Average	
Improving	4 (45%)	2 (22%)	6
Stable	2 (22%)	3 (33%)	5
Worsening	3 (33%)	4 (45%)	7
Total	9 (100%)	9 (100%)	18

tau (τ) = 0.21

second indicator, the two highest ratios (8.82 and 6.56) seen in Churiya-mai and Thulo Sirubari, respectively, have an average forest stock that is improving (Churiyamai) or holding stable (Thulo Sirubari). The two lowest ratios (0.24 and 0.32), in Doramba and Shantipur, are associated with an average stock that is either improving (Doramba) or worsening (Shantipur).

Furthermore, table 8.3 indicates that there is little association between forest condition and population growth for these 18 communities even though they experienced higher growth rates than others in the region. The tau measure of association between the two variables is quite low at 0.24. In locations with above-average population growth, 67 percent of forests are improving or stable in condition. In locations with below-average population growth, 55 percent of forests are improving or stable, while 45 percent are worsening. These data demonstrate that a simple negative relationship between population growth and forest condition does not hold for these 18 cases.

These brief comparisons illustrate a simple point: explanations of forest condition that rely primarily on population pressure may be too simplistic. The entire range of forest conditions can be seen to be associated with high or low values of demographic indicators. Clearly, demographic variables by themselves do not appear to satisfactorily explain forest condition. Two pertinent questions emerge from this finding: (1) how is it that some forests are in better condition in locations where population growth and population density per unit area of forest is high, and

Table 8.4
Preliminary comparisons of forest condition with collective activity

Site Location	Forest Condition Trend	Forest Stock Condition	Collective Activity[a]
Churiyamai (Makwanpur)	Improving	Average	High
Bijulikot (Ramechhap)	Improving	Average	High
Doramba (Ramechhap)	Improving	Average	High
Raniswara (Gorkha)	Improving	Average	High
Bandipur (Tanahun)	Improving	Above average	High
Manichaur (Kathmandu)	Improving	Average	Moderate
Riyale (Kavre Palanchowk)	Stable	Below average	Moderate
Thulo Sirubari (Sindhupalchowk)	Stable	Average	Moderate
Barbote (Ilam)	Stable	Average	Moderate
Baramchi (Sindhupalchowk)	Stable	Below average	Low
Bhedetar (Dhankuta)	Worsening	Above average	Moderate
Agra (Makwanpur)	Worsening	Average	Low
Chhimkeshwari (Tanahun)	Worsening	Average	Low
Chunmang (Dhankuta)	Worsening	Average	Low
Bhagwatisthan (Kavre Palanchowk)	Worsening	Below average	Low
Sunkhani (Nuwakot)	Worsening	Below average	Low
Chhoprak (Gorkha)	Worsening	Below average	None
Shantipur (Ilam)	Worsening	Average	None

a. Organized collective action level at the user level. Low = individuals may observe harvesting constraint on their own, no group activities. Moderate = as a group, individuals have harvesting constraints, minimal group activities, little or no monitoring. High = enforced harvesting constraints, organized group activities, monitoring by members.

(2) how is it that locations with low population growth and density have deteriorating forests?

A look at table 8.4 shows the association of trend in forest condition with a different kind of measure. This measure, called *degree of collective activity,* indicates the extent to which local residents have organized themselves to manage forest use. The degree of collective activity is derived from a set of questions that ask whether there are rules (formal and informal) related to entry into a forest, harvesting in a forest, and monitoring of a forest and how the group organizes its forest-related activities.

A low degree of collective activity is noted for cases in which individuals are aware of forest degradation and resource scarcity and observe harvesting constraints on their own, without any group-level activities or rules of harvest. For this study, I classify low collective activity along with no collective activity. A moderate level of collective activity is noted when a group has harvesting and entry rules, planned minimal forest-related group activities, but little or no monitoring of rule breakers. A high level of collective activity is noted when a group has harvesting and entry rules, monitoring by members, and organized forest-related group activities. These, of course, comprise just a small portion of the repertoire of rules that may exist at any location and are used here as minimum indicators of collective activity. The locations in table 8.4 are arrayed according to the trend in forest condition observed, from improving to worsening.

In table 8.4, five of the six improving forests are associated with high levels of collective activity, while one forest is associated with a moderate level of collective activity by users. All six had stocks that were at least average in condition for this physiographic zone. Four of five forests in stable condition have a moderate level of collective activity associated with them, while one has a low level of collective activity. Three of these stable forests have average stocks and two have below-average stocks. Six of seven forests in worsening condition had low or zero levels of collective activity by villagers, while one forest had villagers engaging in a moderate level of collective activity. Of these seven forests, one had above-average forest stock, three had average forest stocks, and three had below-average forest stocks.

A strong degree of association is evidenced by the tau measure of association for table 8.5. Where a high level of collective activity related to forest management was seen, all forests (100 percent) were improving in condition. There was little or no collective activity being undertaken by the local community in locations where more forests (75 percent) were found to be deteriorating. In the majority of locations where the users were engaged in at least moderate collective action, the forest resource

Table 8.5
Association of level of collective activity with forest condition

Forest Condition	Collective Activity			Total
	High	Moderate	Low or None	
Improving	5 (100%)	1 (20%)	0	6
Stable	0	3 (60%)	2 (25%)	5
Worsening	0	1 (20%)	6 (75%)	7
Total	5 (100%)	5 (100%)	8 (100%)	18

tau (τ) = 0.80

was seen to be neither deteriorating nor improving—that is, forest conditions were stable.

Discussion of Selected Cases

For almost all of the locations in this study, the level of collective activity undertaken by users is found to be positively associated with forest condition. To understand the mechanisms that lie behind these positive associations, this section examines in greater depth two cases for each type of forest trend observed (table 8.6). These cases are selected because they are representative of the larger set in terms of the variance of the factors to be examined and because their case histories provide the most salient detail for the purposes of this study (IFRI, 1995, 1996, 1997a, 1997b).

Improving Forest Conditions

Raniswara   This location is marked by large size, a high level of population growth, and fluctuating migratory patterns. It is also very close to the bustling Gorkha bazaar, the major commercial center in the area. The residents of this VDC have one of the most successful, nationally recognized, active, and well-endowed community forest associations. There are 11 settlements around a large forest (300 ha), with all but two divided along caste lines. There has been no external intervention to speak of in this area; villagers regard the government as a source of neither support nor hindrance.

Table 8.6
Cases selected for discussion

Site Location	Population Growth (percent)	Households per Hectare	Forest Stock	Forest Condition	Collective Activity
Raniswara (Gorkha)	4.71	1.35	Average	Improving	High
Churiyamai (Makawanpur)	5.42	8.82	Average	Improving	High
Riyale (Kavre Palanchowk)	3.00	3.17	Below Average	Stable	Moderate
Barbote (Ilam)	3.64	1.80	Average	Stable	Moderate
Agra (Makawanpur)	0.29	0.37	Average	Worsening	Low
Chunmang (Dhankuta)	4.13	0.68	Average	Worsening	Low

The forest association for this group of users was formed informally seven years ago (with no prior history of organizing in this manner) and legally registered two years later, making it the oldest registered group in the district and one of the oldest in the country. The primary reason for forming the association was to initiate an organized way of protecting a completely denuded hillside—the result of prolonged government neglect, overuse by locals, and land grabbers. In time, the protected area increased, and the association has now petitioned the forest office to add an additional 125 degraded ha to the forest area. In anticipation of a positive response it has initiated planting and protection of seedlings. Forest products are plentiful, but consumption is strictly regulated by the association. Although timber trees are abundant, the annual consumption of timber is being reduced and closely monitored. Very minor infractions take place. Most of the users have switched to using privately grown fodder trees and agricultural residue for their stall-fed cattle, although grass may be cut from the forest floor at all times. Less and less agricultural land is being used for staples because most of the youth labor force is in school. Many farmers are experimenting with fruit trees and vegetables.

This forest association has fashioned several innovative solutions to day-to-day forest-related problems. To deal with political partisanship (which is wrecking many user groups in Nepal), it has banned political discussions in any forum related to this association. To deal with its large numbers (over 2,600 individuals), it has created smaller subcommittees specifically oriented to reducing the load on the executive committee and enhancing the association's ability to cope with large, complex tasks. Users' households are divided along ward lines into subgroups for weeding and protecting the forest area closest to their settlements. To use their time most efficiently in forest-related work, users synchronize weeding, pruning, and coppicing activities with forest-product allocation and distribution activities.

To monitor the use of valuable products such as timber, this association has an investigative subcommittee that monitors the amount requested for a particular use by a user, the amount granted by the association harvest subcommittee, and the ultimate use of the harvested timber by that user. During periods of high usage the association increases the number of forest guards and patrols. To reduce the use of fuelwood, it gives small grants to those who want biogas plants—enough to cover expenses incurred in addition to the available government subsidy.

The association has a regular outreach effort that encourages settlements near the forest borders to join the association or to form their own association. The rationale is that if currently unauthorized users were to become part of the association, costs related to monitoring and sanctioning would decrease, and the pool of labor available for protection and maintenance activities would increase. If unauthorized users form their own association for forest land in their own areas, heretofore unprotected forest lands get protected, and there are fewer occasions of unrestrained harvesting in surrounding forested areas. The Raniswara forest association also regularly sends two trainers to participate in government-sponsored training programs that are held for fledgling forest associations in the region.

Churiyamai   This site is located about 8 km northeast of Hetauda municipality, the center of Makawanpur district, and is accessible by an all-weather road. The three settlements in this site comprise an informal forest association with a total of 750 households and 4,500 individuals. This association has a 19-member executive committee to manage its community forest of about 85 ha. While agricultural production is comparatively low, most residents here have supplementary cash income from selling milk and some poultry. The milk-producing buffalo is stall-fed in all homes. Most of the other livestock is grazed in fields, bunds, and risers. Almost every household has someone working on an off-farm job in neighboring Hetauda or in Kathmandu. Twenty-five percent of the households also have a member working as seasonal labor.

The community forest has two distinct blocks—one of which is a 27-year-old former government research tract and the other a tract initially developed by the Terai Community Forestry Development Program seven or eight years ago. In 1990, the households of the two proximate settlements formed a forest association with a committee to manage both blocks as one community forest. The third settlement disputed this arrangement because the villagers in this settlement were also traditional users and because some parts of the forest were within their boundaries.

As a countermove, this settlement formed a forest association and committee for its own area of the forest. This arrangement was not satisfactory and led to conflicts over boundaries and membership among the three settlements. Resolution to the problem was reached by merging the two groups into one forest association and allowing all three settlements to avail of the entire forest area.

This larger group of users from the three settlements operates on an informal level and is yet to be registered as a forest association under community forestry law. However, they function as a well-organized association, with rules specifying entry, harvest of particular products, and times of harvest. Grazing and felling of live trees is prohibited. Collection of fallen leaves and grass is permitted on payment of a fee. These fees and proceeds from sale of deadwood or fallen trees provide cash income for the association. The income is used to pay for two full-time forest monitors at present. These measures have considerably improved the condition of the forest. The association members also feel that once their application for formal recognition is accepted by the forest office, they will be able to further this improvement by implementing some forest-management, plantation, and erosion control activities that they have planned.

The strict conservation practices have resulted in people planting fodder trees on private land and using a government forest that is almost two hours distant by foot. Residents have also increased their use of agricultural residue and grass from fields and roadsides to supplement animal feed requirements. Like Raniswara, this group has a large repertoire of enforced rules on entry and harvest, and users have high levels of rule awareness and compliance. There are no plans to ease restrictions on cutting of tree fodder or felling of trees.

Stable Forest Conditions

Riyale   Three settlements with a total of 92 households constitute the users of a forest area of 29 ha in this location in Riyale VDC. The forest is within a 20-minute walk of the settlements. There is a market 10 km distant and accessible by a fair-weather road. This VDC is geographically close to Kathmandu valley, but residents have not taken advantage of their location to obtain agricultural inputs or exploit markets for their produce. There is a dairy cooperative nearby that obtains some of its milk supply from the residents of this group.

The forests in this area did not have an organized form of forest protection or management in the past. There was an increasing trend toward degradation until the late 1980s, when mature trees of several valuable timber species were removed. As the forest area deteriorated, villagers started restricting their own harvest of timber as well as any use of their forest by outsiders. The local forest office underwrote a major plantation effort in 1992 and deputed a forest watcher for a period of five years to help monitor the plantation.

This forest association has been able to close the forest to grazing and harvesting of tree products but allows collection of grass and deadwood. There have not been any efforts to raise funds for the association, and besides the initial plantation of saplings, members have not participated in maintenance and protection activities. This is the extent to which they have implemented their management plan. Activities like weeding, thinning, and pruning are planned but yet to be carried out. The presence of a government-paid monitor has reduced illegal activities but not stopped them. There are some violations of the timber harvesting, grazing, and tree fodder rules. However, no fines are levied, and no records are kept of violations.

The forest has not deteriorated since the association was organized in 1991. The general restriction on tree harvest and grazing, and the presence of the forest watcher, has resulted in some regrowth of natural vegetation.

Barbote   Barbote VDC of Ilam district is about a two-hour walk by all-weather road (40 minutes by bus) from Ilam Bazaar. This VDC contains a large forested area (120 ha) that has been looked after by a formally registered forest association for the last six years. There are nine settlements in the immediate surroundings with several others nearby. While the forest in this area did not undergo the rapid deforestation that occurred in central and west Nepal in the 1970s, there was a distinct period of time about eight to 10 years ago when the forest had degraded. The forest improved after villagers started protecting the area. However, in the last three years or so, the forest has begun to show signs of degradation again, and villagers have begun to worry about the future availability of supplies of timber, fuelwood, and fodder.

The community forest boundaries have not been demarcated at any time; a rough estimate was made at the time of the formation of this association. Many members of this association dispute the existing boundaries of the community forest. These members have maintained agricultural plots within, or encroaching on, existing forest land. They hope to claim ownership over these plots if and when the community forest gets demarcated properly.

Population growth is stable with very little fluctuation. Most of the villagers have been here for five or six generations. The executive committee of this association has undergone some upheavals in the past two or three years owing to the resignation—on corruption charges—of the secretary and chair. The users in the immediate vicinity are not very active but do participate in a bare minimum fashion that allows them to remain members.

There are more registered users than actual users: merchants in the nearby market are registered as members but in reality do not use the forest and do not help with any maintenance activities. Villagers point to this membership problem as the reason for the breakdown in cooperation. Falsely registered members outnumber actual members in the register and are able to affect quorum requirements for any change in rules, especially those related to membership. Thus, by their absence they guarantee their membership. When approached by executive committee members to help in the matter, the district forest office has stated that the forest is now a community forest, and, therefore, unless the majority of users complains about a problem, the government can do nothing.

One member acts as the organizer, facilitator, and adviser-at-large for this association. He mobilizes users from time to time for certain activities but now says that it has been getting harder and harder to get the association enthused about the community forest, especially because of the membership and politics problems. As in Riyale, the users in Barbote also have rules constraining entry and harvest, but there is no arrangement for regular monitoring, and there are infractions that are not punished. Because of an ugly history of abuse of authority by office bearers of this association and, now, politics, there is always suspicion among the general body of users about the motives of any activity proposed by an office bearer. There is limited interaction between users, and they rarely assemble in full strength. Decisions requiring general body agreement are not made and, in the case of Barbote, are almost impossible to make because of the difficulty in reaching the quorum requirement.

Worsening Forest Conditions

Agra   This site is within a half-hour walk from a national highway and market. The forest used is about 190 ha and is within a 15-minute walk of the two settlements in the site. Residents of both settlements belong to the same ethnic group and religion and are the traditional users of the forest, although residents of neighboring villages are not barred from harvesting forest products in this forest. For a period of 18 years up to 1989, there was some system of forest protection by the villagers of the locale. In fact, from 1987 to 1989, the users had formed an executive committee to oversee forest-management activities in a formalized manner for the users of the two settlements. In 1990, following political upheaval in the country, this system broke down, and there was no organized form of forest protection or use. Users divided along party lines, and few were willing to reconcile in the matter of resource protection and management. In 1993, villagers from the two settlements again defined a group of users for this forest and elected an executive committee with the objective of preventing tree felling by anyone and of stopping neighboring villages from using the forest. This lasted until 1995 and then again dissolved because there was no agreement over the fines to be levied on rule breakers.

Although there is no organized activity at present, the users of these two settlements have once again defined a user group for this forest, formed an executive committee, and drafted an article of association in preparation for being recognized by the district forest office. The neighboring villagers, however, are opposed to this limited user group and want to be part of it. The main reason these neighbors want to be members appears to be the presence of a slate quarry of 10 to 12 ha that lies within the forest boundary closest to their villages. Several members of those villages have profited from the slate quarry until now, and this important source of income would become off limits once the proposed user group is recognized by the forest office. The application for the forest association is stalled at the forest office because of this opposition, partly because the license for quarrying slate was issued by the district development committee office, a higher-level authority.

Villagers of the two proximate settlements have appealed to the district soil conservation office to stop the slate mining because large-scale erosion is taking place at the site. The erosion gullies and runoff are destroying vegetation in the immediate forest area. In the meantime, valuable herbs are being harvested indiscriminately and sold to outside contractors, and unrestrained grazing and cutting of fodder takes place.

Chunmang   The site in this VDC is not very accessible: a steep downhill walk of three hours from the road head, Hile (at 2,300 m), gets one to the site (between 600 and 900 m). The nine settlements in this location are scattered on the west-facing slope of a mountain, six settlements are closer to the area's forest, and three settlements are farther away. All the settlements are situated higher than the forest area, which ends at the streambeds along the base of the mountain. The residents of this site live in settlements differentiated mainly along caste lines; all castes are present. One particular caste is dominant, politically and socioeconomically, by virtue of their numbers. The local representative to the political party in power is from this caste. They also have a loyal following of some members of lower caste, who depend on them for employment and land.

There has been discord over organizing these settlements to manage the nearby forest in the past several years, owing mostly to the various hindrances put up by the dominant caste. Of the nine settlements using this forested area over the last several decades, there is divided opinion over the options for managing the forest area. The users have been discussing variations of two options: (1) to combine all nine settlements and form one association and one large forest area with different management units or (2) to form two associations and split up the forest area according to relative distance to forest from settlements. Of the six settlements that are closer to the forest, two (led by the dominant caste) are unwilling to form a large association that combines both far and near settlements and utilizes the entire forest area. Their first proposal is to have one portion (the larger, more valuable forest) allocated to the six settlements and another portion (the smaller, more degraded) allocated to the three distant settlements, thus forming two associations with two separate areas. Their second proposal is simply to exclude the three distant settlements and form one association for the entire forest area. Neither option is acceptable to the three settlements because they see the allocation of forest area as unfair in the first case and their complete exclusion from forest use as an insult to their traditional rights in the second case.

The opposition put up by the dominant caste members in one of the six proximate settlements has been frustrating to the more cooperative villagers who belong to other castes in these six settlements, especially because the forest is currently open to anyone for use. As a result, many areas in the forest are getting degraded, with other areas soon to follow. Most of these villagers are willing to form a single association with the three distant settlements or even participate in an equitable apportioning of the forest land to two associations. Without some form of collective action, all agree, there will be problems in the near future with regard to forest products.

This situation has also been frustrating for the staff of the district forest office, who tried about four years ago to establish an association but were rebuffed in their efforts by the dominant caste. Since then, however, there has been no attempt by anyone outside these communities to try again. There are several individuals in and around the area who would like to assist in forming an association for this forest, but these individuals say that they would like a third party to act as an intermediary to mediate and give advice on other options for all these forest users. In the meantime, the forest is a source of timber, fodder, and fuel for all these settlements and even for some outsiders.

As in Agra (and Barbote), district officials have failed to act on petitions in Chunmang. This lack of action has created uncertainty for the users and has helped opportunistic individuals take advantage of the lack of any organized form of forest protection by harvesting timber and encroaching on forest land. In both Agra and Chunmang, villagers are aware of the deteriorating condition of their forest resources, but no group activity is evident, partly because of factionalization of the community owing to politics and economic ties. However, there was a time in both locations when some form of organized activity had started and subsequently failed; both locations have had group-building efforts by outside agencies four or five years in the past, but none are going on at present.

Conclusion

This study examined the relationship between population, institutions, and forest conditions in the Middle Hills of Nepal. The study indicated that the variation in population growth rates across the locations studied had almost no discernible correlation with the variation of forest condition in those locations. The study did, however, show a strong association between local collective action and variation in forest conditions across the 18 cases.

By identifying some of the characteristics of institutional arrangements used by villagers, this study sought to appraise an undervalued facet of the complex presentation of the population-environment dynamic. That local forest users can cope with perceived changes in resource condition and in user population is evident from the cases studied in this chapter. In the more successful cases, arrangements for identifying genuine users, determining harvest amounts and timing, and active monitoring by users themselves emerge as important factors in managing forest resources (table 8.7).

Table 8.7
Some institutional characteristics of select cases

| | | Institutional Characteristics | | |
Site Location	Forest Condition	Entry and Harvest Restrictions	Monitoring Arrangements	Adaptive or Innovative Mechanisms
Raniswara (Gorkha)	Improving	Yes	Yes	Yes
Churiyamai (Makawanpur)	Improving	Yes	Yes	Yes
Riyale (Kavre Palanchowk)	Stable	Yes	Yes	No
Barbote (Ilam)	Stable	Yes	No	No
Agra (Makawanpur)	Worsening	No	No	No
Chunmang (Dhankuta)	Worsening	No	No	No

Where users were unable to define the extent of forest boundaries or the number of users in a group clearly, the ambiguity allowed opportunistic individuals to encroach on forested land. Investments in monitoring, in particular, significantly determine the difference between a flourishing resource and one just able to meet the needs of users. In the locations with higher populations but improving resources, Raniswara and Churiyamai, user groups invested in monitoring, even to the point that extra guards were assigned during seasons of greater need. This finding follows a study by Agrawal and Yadama (1997), who, in their sample of 279 communities, found that the most important form of user participation was the level of investment by the user group in monitoring and protecting activities.

Much of the literature on collective action has discussed the negative association between group size and collective action. Yet in groups such as in Raniswara, users had ways to deal with large numbers. The adaptation of user-group structure by creating levels of subgroup activity was one way to deal with the increased complexity of tasks and the difficulty of coordination that is brought on by large memberships. This sort of innovation was facilitated at times by the village administration and forestry officials who participated in the meetings that assign duties and responsibilities to various subgroups.

The group in Raniswara has also actively pursued the objective of increasing the area of forest it uses by soliciting the membership of neighboring villages, which then attach their adjacent forest lands to that of the group. Arranging for regular interactions between users, other villagers, and external parties in positions of authority and influence had the effect of reducing suspicion, facilitating information diffusion, raising awareness throughout the area, and garnering public support for management and conservation ideas. A breakdown in community relations and an undermining of collective organization and action was seen in Barbote, Agra, and Chunmang, where the public was divided in its opinion (due to kinship, economic ties, allegations of corruption, and politics) and no third party was available (or interested) to mediate the conflict.

The World Bank has stated that "because the people who cut or plant trees typically *have no incentive* [emphasis added] for considering the environmental and social consequences of their actions, externalities inexorably lead to excessive deforestation and insufficient planting of new trees" (World Bank, 1991, 9). Such statements have been acted on in the past with the result that disproportionately large funds have been allocated to reforestation and strengthening the administrative functioning of government forest offices. However, the findings of this study suggest a different direction and point of emphasis in policy research and application. The recognition of the mediating effects of local institutional arrangements in the population-environment dynamic has important ramifications for those who seek to support community forestry and, more generally, participatory approaches to governing natural resources. This study suggests that development policy aimed at preserving the environment must recognize the significance of institutional arrangements at the local level to resource conditions at that level. Ultimately, the benefits and costs associated with resource conditions at the local level have considerable bearing on larger environmental issues. Furthermore, the study suggests that government policy on participatory resource management will be more successful if it is facilitative of institutional innovation and adaptation at the village level.

Acknowledgments

I gratefully acknowledge the support of the Forests, Trees, and People Programme of the Food and Agriculture Organization of the United Nations, the United Nations Population Fund (INT/94/040), the Hills Leasehold Forestry and Forage Development Project of His Majesty's Government of Nepal/FAO/IFAD, the MacArthur Foundation, and the Workshop in Political Theory and Policy Analysis at Indiana University. This chapter is the first of a series of comparative studies of forest institutions drawing on the International Forestry Resources and Institutions research program in Nepal. I thank the members of the program team, especially Parsu Ram Acharya, Sudil Gopal Acharya, Mukunda Karmacharya, and Birendra Karna, for their diligence and support during fieldwork, and the Nepali villagers who were generous and patient with our inquiries during visits to their locations. I am grateful to Elinor Ostrom and Clark Gibson for detailed comments on several drafts of this chapter.

I also acknowledge Keshav Kanel, Paul Turner, Robin Humphrey, and Neeraj Joshi for their help at various stages of preparing this chapter.

Note

1. Varughese (1999) sampled six of the 18 cases to examine the change in forest condition after a period of four years. Tree, sapling, and shrub species were counted using stratified random sampling during revisits to the sites. Five of the six sites returned stem counts, girth, and species richness that validate the perceptions of the residents and foresters. The sixth case differed only on the density for shrub species.

References

Abernathy, Virginia. 1993. *Population Politics: The Choices That Shape Our Future*. New York: Plenum/Insight.

Acheson, James. 1989. "Where Have All the Exploitings Gone? Co-management of the Maine Lobster Industry." In *Common Property Resources,* ed. Fikret Berkes, 199–217. London: Belhaven.

Agrawal, Arun, and Gautam Yadama. 1997. "How Do Local Institutions Mediate Market and Population Pressures on Resources? Forest Panchayats in Kumaon, India." *Development and Change* 28: 435–65. Oxford: Blackwell.

Arizpe, Lourdes, M. Stone, and D. Major, eds. 1994. *Population and Environment: Rethinking the Debate*. Boulder, CO: Westview Press.

Arnold, J. E. M., and J. G. Campbell. 1986. "Collective Management of Hill Forests in Nepal: The Community Forestry Development Project." In *Proceedings of the Conference on Common Property Resource Management,* National Research Council, 425–54. Washington, DC: National Academy Press.

Bajracharya, D. 1983. "Deforestation in the Food/Fuel Context: Historical and Political Perspectives from Nepal." *Mountain Research and Development* 3: 227–40.

Becker, C. Dustin, Abwoli Y. Banana, and William Gombya-Ssembajjwe. 1995. "Early Detection of Tropical Forest Degradation: An IFRI Pilot Study in Uganda." *Environmental Conservation* 22(1) (Spring): 31–38.

Bilsborrow, Richard, and P. DeLargy. 1991. "Land Use, Migration, and Natural Resource Degradation: The Experience of Guatemala and Sudan." In *Resources, Environment, and Population: Present Knowledge, Future Options,* ed. K. Davis and M. Bernstam. New York: Oxford University Press.

Binswanger, Hans, and P. Pingali. 1989. "Population Growth and Technological Change in Agriculture." In *Population and Resources in a Changing World,* ed. K. Davis and M. Bernstam. Stanford, CA: Morrison Institute for Population and Resource Studies.

Boserup, Esther. 1965. *The Conditions of Agricultural Growth: The Economics of Agrarian Change under Population Pressure.* London: Allen & Unwin.

———. 1981. *Population and Technological Change.* Chicago: University of Chicago Press.

Brown, Lester, E. Wolf, and L. Starke, eds. 1987. *State of the World 1987.* New York: Norton.

Central Bureau of Statistics (CBS). 1995. *Population Monograph of Nepal.* Kathmandu: National Planning Commission Secretariat.

———. 1996. *Nepal Living Standards Survey.* Kathmandu: National Planning Commission Secretariat.

Chakraborty, Rabindra N., Ines Frier, Friederike Kegel, and Martina Mascher. 1997. *Community Forestry in the Terai Region of Nepal: Policy Issues, Experience, and Potential.* Working Paper 5. Berlin: German Development Institute.

Chhetri, R. B., and T. R. Pandey. 1992. *User Group Forestry in the Far Western Region of Nepal.* Kathmandu: ICIMOD.

Cruz, Maria C., Carrie Meyer, Robert Repetto, and Richard Woodward. 1992. *Population Growth, Poverty, and Environmental Stress: Frontier Migration in the Philippines and Costa Rica.* World Resources Institute Report. Washington, DC: World Resources Institute.

Dahal, D. R. 1994. *A Review of Forest User Groups: Case Studies from Eastern Nepal.* Kathmandu: ICIMOD.

Eckholm, E. P. 1975. "The Deterioration of Mountain Environments." *Science* 189: 764–70.

———. 1976. *Losing Ground: Environmental Stress and World Food Prospects.* New York: Norton.

Ehrlich, Paul, and A. Ehrlich. 1991. *The Population Explosion.* New York: Touchstone, Simon & Schuster.

Exo, S. 1990. "Local Resource Management in Nepal: Limitations and Prospects." *Mountain Research and Development* 10(1): 16–22.

Feeny, David, Fikret Berkes, Bonnie McCay, and James Acheson. 1990. "The Tragedy of the Commons: Twenty-two Years Later." *Human Ecology* 18(1): 1–19.

Fox, Jefferson. 1993. "Forest Resources in a Nepali Village in 1980 and 1990: The Positive Influence of Population Growth." *Mountain Research and Development* 13(1): 89–98.

Gilmour, D. A., and R. J. Fisher. 1992. *Villagers, Forests, and Foresters: The Philosophy, Process, and Practice of Community Forestry in Nepal.* Kathmandu: Sahayogi Press.

Gilmour, D. A., and M. Nurse. 1991. "Farmer Initiatives in Increasing Tree Cover in Central Nepal." In *Workshop on Socioeconomic Aspects of Tree Growing by Farmers.* Anand, Gujarat, India: Institute of Rural Management.

Griffin, David M. 1988. *Innocents Abroad in the Forests of Nepal.* Canberra, Australia: Anutech.

Gronow, Jane, and N. K. Shrestha. 1991. *From Mistrust to Participation: The Creation of a Participatory Environment for Community Forestry in Nepal.* Social Forestry Network Paper 12b. London: Overseas Development Institute.

HMG (His Majesty's Government). 1988. "Master Plan for the Forestry Sector, Nepal." Main Report in *Master Plan for the Forestry Sector Project.* Kathmandu: Ministry of Forests and Soil Conservation.

Hobley, M. E. A. 1990. "Social Reality, Social Forestry: The Case of Two Nepalese Panchayats." Ph.D. dissertation, Australian National University.

International Forestry Resources and Institutions (IFRI) Reports. 1995. *Site Reports for Churiyamai, Baramchi, Riyale, and Bijulikot.* Kathmandu: IFRI-Nepal and Hills Leasehold Forestry and Forage Development Project.

———. 1996. *Site Reports for Manichaur and Sunkhani.* Kathmandu: IFRI-Nepal and Shivapuri Integrated Watershed Development Project.

———. 1997a. *Revised Site Reports for Thulo Sirubari, Doramba, Agra, and Bhagawatisthan.* Kathmandu: IFRI-Nepal and Hills Leasehold Forestry and Forage Development Project.

———. 1997b. *Site Reports for Chhimkeshwari, Chhoprak, Raniswara, Bandipur, Barbote, Shantipur, Chunmang, and Bhedetar.* Kathmandu: IFRI-Nepal.

Ives, Jack D. 1987. "The Theory of Himalayan Environmental Degradation: Its Validity and Application Challenged by Recent Research." *Mountain Research and Development* 7: 189–99.

Ives, Jack D., and B. Messerli. 1989. *The Himalayan Dilemma: Reconciling Development and Conservation.* London: Routledge.

Jolly, Carole. 1994. "Four Theories of Population Change and the Environment." *Population and Environment* 16(1): 61–90.

Karki, M., J. B. S. Karki, and N. Karki. 1994. *Sustainable Management of Common Forest Resources: An Evaluation of Selected Forest User Groups in Western Nepal.* Kathmandu: ICIMOD.

Mahat, T. B. S., D. M. Griffin, and K. R. Shepherd. 1986a. "Human Impact on Some Forests of the Middle Hills of Nepal. Part 1. Forestry in the Context of the Traditional Resources of the State." *Mountain Research and Development* 6: 223–334.

———. 1986b. "Human Impact on Some Forests of the Middle Hills of Nepal. Part 2. Some Major Human Impacts before 1950 on the Forests of Sindhu Palchowk and Kabhre Palanchowk." *Mountain Research and Development* 6: 325–34.

———. 1987a. "Human Impact on Some Forests of the Middle Hills of Nepal. Part 3. Forests in the Subsistence Economy of Sindhu Palchowk and Kabhre Palanchowk." *Mountain Research and Development* 7(1): 53–70.

————. 1987b. "Human Impact on Some Forests of the Middle Hills of Nepal. Part 4. A Detailed Study in Southeast Sindhu Palchowk." *Mountain Research and Development* 7(2): 111–34.

Messerschmidt, D. A. 1986. "People and Resources in Nepal: Customary Resource Management Systems of the Upper Kali Gandaki." In *Proceedings of the Conference on Common Property Resource Management,* 455–80. Washington, DC: National Academy Press.

Metz, John J. 1990. "Forest-Product Use in Upland Nepal." *Geographical Review* 80(3) (July): 279–87.

Myers, Norman. 1991. "The World's Forests and Human Populations: The Environmental Interconnections." In *Resources, Environment, and Population: Present Knowledge, Future Options,* ed. K. Davis and M. Bernstam. New York: Oxford University Press.

Netting, Robert. 1993. *Smallholders, Householders: Farm Families and the Ecology of Intensive, Sustainable Agriculture.* Stanford, CA: Stanford University Press.

New ERA. 1996. *Community Forest Management with Relation to Population.* Discussion Paper. Kathmandu: New ERA.

Ostrom, Elinor. 1998a. "The International Forestry Resources and Institutions Research Program: A Methodology for Relating Human Incentives and Actions on Forest Cover Biodiversity." In *Forest Biodiversity in North, Central and South America and the Caribbean: Research and Monitoring,* ed. F. Dallmeier and J.A. Comiskey, Man and the Biosphere Series, vol. 22, 1–28. Paris: UNESCO.

————. 1998b. "Self-Governance of Common-Pool Resources." In *The New Palgrave Dictionary of Economics and the Law,* vol. 3, ed. Peter Newman, 424–33. London: Macmillan.

————. 1990. *Governing the Commons: The Evolution of Institutions for Collective Action.* New York: Cambridge University Press.

Pradhan, Ajay P., and Peter J. Parks. 1995. "Environmental and Socioeconomic Linkages of Deforestation and Forest Land Use in the Nepal Himalaya." In *Property Rights in a Social and Ecological Context,* ed. Susan Hanna and Mohan Munasinghe. Washington, DC: Beijer International Institute of Ecological Economics and World Bank.

Shivakoti, Ganesh, G. Varughese, E. Ostrom, A. Shukla, and G. Thapa, eds. 1997. *People and Participation in Sustainable Development.* Proceedings of an International Conference. Bloomington: Indiana University, Workshop in Political Theory and Policy Analysis; and Rampur, Chitwan, Nepal: Tribhuvan University, Institute of Agriculture and Animal Science.

Simon, Julian. 1981. *The Ultimate Resource.* Princeton, NJ: Princeton University Press.

————. 1983. "Population Pressure on the Land: Analysis of Trends Past and Future." *World Development* 11(9): 825–34.

———. 1990. *Population Matters: People, Resources, Environment and Integration.* New Brunswick, NJ: Transaction.

Subedi, Bhim P. 1997. "Population and Environmental Interrelationships: The Case of Nepal." In *Population, Environment, and Development,* ed. R. K. Pachauri and L. F. Qureshy. Proceedings of an International Conference. New Delhi: Tata Energy and Resources Institute.

Varughese, George. 1999. "Villagers, Bureaucrats, and Forests in Nepal: Designing Governance for a Complex Resource." Ph.D. dissertation, Indiana University, Bloomington.

Wilson, E. O. 1992. *The Diversity of Life.* New York: Norton.

World Bank. 1978. *Forestry.* Sector Policy Paper. Washington, DC: World Bank.

———. 1991. *The Forest Sector.* World Bank Policy Paper. Washington, DC: World Bank.

Forests, People, and Governance: Some Initial Theoretical Lessons

Clark C. Gibson, Elinor Ostrom, and Margaret A. McKean

The authors of this volume began their empirical analyses from a shared conceptual foundation and a common suite of measures. The International Forestry Resources and Institutions (IFRI) research program, which was built on the Institutional Analysis and Development framework, provided this starting point: an approach with methods to measure systematically a number of social and biophysical factors important to explaining the interaction of communities and their forest resources (see the appendix to this volume). The multiple variables included within the IFRI protocols—including those that might be investigated by anthropologists, economists, political scientists, foresters, ecologists, sociologists, and lawyers—allow for the testing of a large number of theories using IFRI data.

This volume attests to this breadth. The authors examined theories and topics regarding the nature of property rights, collective action, rule enforcement, human foraging patterns, markets, transportation systems, informal norms, institutional creation and change, ethnicity, agricultural livelihoods, and population. Given the breadth of the questions addressed by the authors, it is now time to ask whether any initial theoretical lessons can be derived from this set of studies.

We think there are. First, the major lesson from these studies is that local users of forest resources can exercise more control over the incentives they face than is frequently depicted in textbooks on natural resource policies. As McKean argues in chapter 2, forest users are *not* trapped in an inevitable race to cut down forests. Many of the forest users in Kumaon, in Nepal, and in Bolivia have spent hours debating with one another over the sources of forest deterioration they perceived and proposing alternative rules for consideration by others in their community. Over time, they experimented with rules relating to who could use local forests, which and when forest products would be harvested, what harvesting tools would be allowed, how the forest could be guarded, and what sanctions would be imposed on those who broke community rules. As a result, many of these forest users successfully overcame the dilemmas they faced to achieve a form of self-government that enabled them to manage local forests better than some of their neighbors and many government or private forests in their countries. Thus, many of these cases demonstrate that common property can be an efficient form of property rights in relationship to common-pool resources, as McKean reasons, rather than being the source of inefficiency, as is still argued in many resource policy textbooks and policy papers (see, for example, World Bank, 1991).

The cases in this volume thus enable us to argue strongly against the widely held presumption that the users of natural resources are always helpless and cannot themselves do anything about resource degradation. The theoretical arguments presented by McKean in chapter 2 are supported by the successful cases that tend to follow design principles such as clearly demarcating boundaries, devising equitable rules for sharing benefits and costs, establishing effective monitoring arrangements for imposing graduated sanctions, and creating larger organizations by nesting smaller units within the larger organizations.

On the other hand, we have learned once again that local actions vary substantially. Members of some communities fail to perceive the growing scarcity of their local forests, fail to create effective rules to counteract the incentives to overharvest, and fail to enforce their own rules. Forest users in Loma Alta, Ecuador, and some of the communities in India and Nepal did not succeed in designing and implementing local rules that effectively controlled the quantity of forest products removed from local forests. Given these recurring differences in the effectiveness of local organization, important questions arise for policy analysts, as well as those interested more generally in the theory of collective action: What factors help to account for these differences among communities in their capacities to design, alter, and implement successful self-governing institutions? Why do some communities self-organize in the first place? Why do some of these continue to experiment with new rules over time so that they achieve relatively efficient as well as sustainable results? Why do some communities neglect to alter organizational designs that, while once successful, fail when external environments change? These are extremely difficult questions to answer. Many combinations of variables affect the initial establishment of new institutional arrangements as well as the effort to adapt and experiment with rules so as to find the right set of incentives given the ecological, cultural, and broader institutional environment involved. As we have been working together to build the IFRI research network and the groundwork for the studies reported herein (as well in conducting other studies to be reported in future publications), we have also begun to develop a better theoretical understanding of the diversity of factors that appear to affect decisions made by local users about whether or not to invest (or, continue to invest) in collective action. It must be stressed that these investments are costly. They involve users engaged in long and sometimes heated debates about whether they are overharvesting, who is to blame, whether rules can be changed, whether the rules used by neighboring communities are better than the ones used locally, and how to enforce these rules.

At one level, providing a theoretical explanation is simple. If local users do not expect that the *benefits* they will receive (in terms of a more sustainable yield, more diverse forest products, reduced erosion, or a better water supply) from designing more effective local institutions will exceed the up-front as well as the continuing *costs* of day-to-day management of a forest, they will not invest in improving their local institutions. Local institutions affect the probability of participants free riding on the input efforts of others and increase the likelihood of positive benefits. The same institutions, however, are costly to design and sustain. So the crucial question for members of a community of resource users is whether the benefits of organizing are worth the costs. Explaining why some communities effectively self-organize while others do not requires that we understand the benefits and costs of self-organization as perceived by diverse members of local communities. Thus, a theoretical grasp of local collective action depends on linking the costs and benefits of investments in local institutions to the decisions made within collective-choice arrangements within a community (see Ostrom, forthcoming, for a more formal and extended presentation).

An abstract theory of the benefits and costs of local collective action is an important part of explaining why some users do and others do not overcome the resource dilemmas they face. A more practical question is which empirical variables affect these benefits and costs. Now the task becomes much tougher. Many variables potentially affect either the benefits or the costs of collective action. Trying to identify these is an important task for policy analysts. If empirical relationships can be established, it may then be feasible to design public policies that reduce some of the costs and increase some of the benefits so that more local users successfully overcome the resource dilemmas they face. Based on the work of many scholars (McKean, chapter 2 this volume; Wade, 1994; Schlager, 1990; Tang, 1992; Ostrom, 1990, 1992a, 1992b; Baland and Platteau, 1996; Ostrom, Gardner, and Walker, 1994), it is possible to begin to identify factors that multiple scholars have identified as enhancing the likelihood that forest-resource users will organize themselves in the first place, and continue to experiment with revised rules, to avoid the social losses associated with ineffective rules-in-use related to the use of a common-pool resource. Ostrom (forthcoming) divides these factors into two sets. The first set refers to the attributes of a resource; the second refers to the attributes of the users of that resource.

Attributes of the Resource:

R1. Feasible improvement: The forest is not perceived to be at a point of deterioration such that it is useless to organize or so underutilized that little advantage results from organizing.

R2. Indicators: The change in quality and quantity of forest products provides reliable and valid information about the general condition of the forest.

R3. Predictability: The availability of forest products is relatively predictable.

R4. Spatial location, terrain, and extent: The forest is sufficiently small, given the terrain, the transportation available, and the communication technology in use, that users can develop accurate knowledge of external boundaries and internal microenvironments, and they can develop low-cost monitoring arrangements.

Attributes of the Users:

A1. Salience: Users are dependent on the forest for a major portion of their livelihood (or for other variables of importance to them).

A2. Common understanding: Users have a shared image of the forest (attributes R1, R2, R3, and R4 above) and how their actions affect each other and the forest.

A3. Discount rate: Most users have a sufficiently low discount rate in relation to future benefits to be achieved from the forest.

A4. Trust and reciprocity: Users trust one another to keep promises and relate to one another with reciprocity.

A5. Autonomy: Users are able to determine access and harvesting rules without external authorities countermanding them.

A6. Prior organizational experience and local leadership: Appropriators have learned at least minimal skills of organization and leadership through participation in other local associations or learning about ways that neighboring groups have organized.

Both sets of variables can affect the costs and benefits of individuals who must decide whether to invest their own resources into constructing or improving a local institution related to their forest. If local users did not expect their forest to improve even with a successful collective effort by individuals (R1), for example, it is highly unlikely that they would organize in the first place or invest in efforts to improve their rules. The unpredictability of growth patterns in a forest (R3) makes it more costly for anyone to figure out effective rules limiting harvests to a sustainable yield. If users do not trust one another to keep promises (A4), they have to expect to pay much higher enforcement costs, which may use up some or all of the benefits they could achieve. Ostrom (forthcoming) has discussed at some length how these variables interact to affect the benefits and costs of local collective action.

In addition to the above variables for which there is a relatively clear theoretical linkage between the variable and the costs and benefits as perceived by users, there are two additional attributes of the users for which there is considerable theoretical dispute. These are the size of the group and the heterogeneity of the users. Theoretical arguments have been made based on the foundational work of Mancur Olson (1965) that smaller groups face lower transaction costs and are thus more likely to overcome collective action problems than larger groups (see also Buchanan and Tullock, 1962; Baland and Platteau, 1996; Cernea, 1989; but see Hardin, 1982). But as we have seen from Agrawal's chapter, smaller groups may be disadvantaged when it comes to marshaling resources sufficient to monitor the use of a forest or to enforce local rules through the use of the courts. We also see a lack of a strong relationship in studies of self-governed irrigation systems. In his study of 37 farmer-governed irrigation systems, Tang (1992) did not find any statistical relationship between the size of the group and performance variables. In Lam's (1998) analysis of a much larger set of irrigation systems in Nepal ranging in size up to 475 irrigators, he did not find a significant relationship between the number of farmers and performance variables. In chapter 8 in this volume where the number of user households ranged from 79 to 750, Varughese found that the number of households per hectare of forest area did not make a systematic difference in the organization or performance of collective action. Consequently, empirical studies are challenging the presumption that smaller groups are more likely to self-organize and be successful in their organization.

The reason that some scholars argue that size is negatively related to the likelihood that users will overcome dilemmas to self-organize to manage a common-pool resource is that they presume that larger groups are also more likely to be heterogeneous than smaller groups. Thus, heterogeneity itself is frequently considered a detriment to self-organization. On the other hand, Mancur Olson (1965) recognized the possibility that groups where considerable heterogeneities exist may be privileged if those with the most economic interests and power were to initiate collective action to protect their own interests. Those with fewer assets might find themselves able to free-ride on the contributions of those with many assets. This argument has been presented in a more rigorous manner by Bergstrom, Blume, and Varian (1986) and given modest support in an experimental setting by Chan et al. (1996). On the other hand, Dayton-Johnson and Bardhan (1998) argue that inequality of assets may be conducive to successful organization within a narrow range but harmful over a broader range. An empirical study by Molinas (1998) supports the notion that the relationship between income inequality and effectiveness of local groups is curvilinear.

Unfortunately, for those who prefer simple explanations of social behavior, this is a large set of variables—12 in all—that potentially affect the cost-benefit calculus of resource users. And, to make things worse, beyond these 12 variables are a large number of other variables identified in the policy literature some scholars presume affect rates of deforestation. These include such popular explanations as population density, availability of new transportation linkages, availability of substitutes for forest products, and increases in the value of timber or other forest products.

Many of the above variables are strongly affected by the larger governmental regimes in which forests are located. National governments can facilitate local self-organization by providing accurate information about natural resource systems, providing arenas in which participants can engage in discovery and conflict-resolution processes, and providing mechanisms to back up local monitoring and sanctioning efforts. The formation of pro-grassroots coalitions of nongovernmental organizations, international donors, and sympathetic political elites makes a major difference in how local users may be able to organize themselves effectively (Silva, 1994; Blair, 1996). Thus, forest users in macropolitical regimes that facilitate their efforts are more likely to develop successful local institutions than those living in regimes that ignore resource problems entirely or, at the other extreme, presume that all decisions about governance and management need to be made by national governments. When rules are imposed by outsiders without consulting those who are most affected, local users are more likely to become robbers, rather than cops, toward the resources they might otherwise have managed sustainably and to try to evade apprehension by the external authorities' cops.

While the concepts of benefits and costs are relatively simple, no single variable (or even two or three variables) is sufficient to provide a firm empirical link to these theoretical concepts. When all relevant benefits and costs can be denominated in currency, then the task of operationalizing and testing a theory is much simpler than when many nonmonetized variables affect the benefit-cost calculus of participants. Further, testing the relative importance of more than a dozen different variables on the likelihood of local users organizing and reorganizing themselves so as to solve resource dilemma problems is not something that can be done with the series of cases presented in this volume. A thorough exploration and testing will require a much larger number of individual studies. Obtaining better measures of these concepts and examining their relative importance in explaining the emergence and sustainability of local organizations for managing forests is a high priority in our current and future research. And the creation of the IFRI research network is designed exactly to enable us to develop a much larger data set for this kind of rigorous comparative analysis. But the cases in this volume provide some added confidence that many of these variables will prove to be important links between the complex environment in which users live and must make difficult choices and the abstract concepts of the costs and benefits of collective action. Let us illustrate how some of these concepts are addressed in many of the chapters of this volume.

In Agrawal's study, for example (chapter 3), forest users in Kumaon had to perceive a feasible improvement in the conditions of the forests (R1) they attempted to manage since they were not in any way forced to organize. As table 3.4 shows, all of the forests do at least have some woody biomass present, and some have a significant level of biodiversity present. Officials of the forest councils do take periodic assessments of the conditions of a forest before deciding on harvesting levels each year (R2). This is especially true of the more successful councils (Agrawal, 1994). Fresh evidence of illegal harvest of forest products can result in the firing of the guard that the council hires. The predictability of forest products (R3), while relatively high, does not vary much among the forests in Kumaon and thus does not play much role in explaining the differences among communities. In Agrawal's study, the smaller communities that were trying to manage a spatially disperse forest (R4) were the ones facing the greatest difficulty in developing low-cost monitoring arrangements over long distances.

As Agrawal points out in figure 3.1 of his chapter, forests do play a critical role in the production activities of Kumaon villagers (A1). Without the fodder, fertilizers, firewood, and construction timber available from forests, often at low cost, hill villagers would find it extremely hard to subsist in this environment. And all of the communities do have the authority to establish village councils with considerable autonomy (A5). Most of the communities share the remaining attributes considered as conducive to self-organization. In fact, all of the communities that Agrawal studied are actually organized and functioning to some extent. What his study adds to our understanding is the importance of matching group size to ecosystem size in overcoming monitoring problems. A group needs to be large enough to mobilize sufficient resources for an effective monitoring program to challenge those who succumb to the temptation to break community rules and eventually to sanction them. If a small, relatively poor village has a relatively large forest to patrol, it faces much higher continuing costs of monitoring and enforcement than a larger village with similar levels of household income that is at least able to mobilize a bigger labor pool for monitoring who is using a forest. Thus, Agrawal's study helps us understand how the size of user groups may have a curvilinear relationship to the likelihood of successful self-organization.

While Banana and Gombya-Ssembajjwe (chapter 4) do not discuss the self-governance of common-property forests, the lessons that emerge from their chapter still highlight the importance of the attributes discussed above. One obvious but important implication is that the spatial location, extent, and terrain of forests (R4) is critical to the construction of successful management regimes, whether management is undertaken by local groups or a central government.

The Ugandan government reserve forests of Lwamunda, Mbale, and Bukaleba are large, with long borders along nongovernment lands. To enforce the country's laws that limit the harvesting of forest products from these reserves, the forest department would need a large number of personnel and, ideally, better means of transportation. Since the government is unwilling to fund the department at such a level to make this possible, people routinely exploit these government forests. Their sheer size helps to transform them into open-access resources. The terrain of the Echuya reserve, on the other hand, helps the department with enforcement by lowering the costs of patrolling. Although large, this forest has only one road nearby, which can be easily monitored.

The attributes that affect decision making can also help illuminate why the private forest of Namungo is not overexploited. With small, short borders and a path around it, Namungo's family and staff can easily patrol the forest's boundaries. Namungo also avoids the costs associated with attributes of the users found in common property situations. As a single owner, he does not have to overcome the differences between himself and other coowners regarding salience (A1), common understanding (A2), or trust (A5). Individual ownership may make other attributes more important, such as discount rate (A3): if Namungo's discount rate changes, he can choose to clear his forest without enduring the costs associated with making a decision through a group.

As in the Banana and Gombya-Ssembajjwe chapter, Schweik's analysis (chapter 5) demonstrates that the attributes of the forest affect its management. The Shaktikhor area's government forests are extensive but sit on difficult terrain with limited transportation infrastructure. This location makes monitoring by personnel of the Department of Forests costly and difficult, resulting in a pattern of enforcement that focuses only on areas that are easily accessed by vehicles.

Some attributes of the users of these forests augur well for the emergence of successful local institutions. All the individuals in the area depend on the forests for their livelihoods, including fodder, fuel, food products, and timber. Most individuals also share common understandings about the role and use of the forests, even though these uses may be unequal (based on the caste system). The villagers perceive the forests' deterioration and recognize the need to do something about it. But without the autonomy to construct rules about the forests' use, locals have not sought to invest in constructing forest-management institutions. The district forest officer does not want to relinquish control over harvesting rules to the local communities.

In the case of Loma Alta, Ecuador, explored by Gibson and Becker (chapter 6) we find many of the user attributes that would reduce the costs of local self-organization. Significant among these are that the community has full local autonomy as well as extensive prior organizational experience. In fact, the community has organized itself for the provision of many local public goods.

The location of the forest, however, has contributed to two problems that have prevented Loma Alta from devising and enforcing rules about its overexploited forest. One has to do with a perception of the extent of the forest. Many Loma Alta residents do not regularly make the long trip to harvest from their forest, and so they continue to imagine that the forest extends much further than it does. These members do not share a common understanding of the problems the forest faces due to the incursion of neighboring users, their own overharvesting, and the link between the forest and their water supply. Second, the large distance between the community and the forest also raises the cost of any effort to try to monitor the use of the forest.

The number of groups in Loma Alta that use the forest for different reasons also poses high costs to anyone seeking to construct effective forest-management plans. The view of the timber cutters differed from those of the paja toquilla farmers and hunters. And few of these groups understand their impact on the forest's condition. Thus, a common understanding about the forest was relatively absent in the Loma Alta case.

It is interesting to note that after our initial research visit, Becker returned to Loma Alta as part of an effort organized by a local NGO to help the local community establish a reserve in their valuable forest (Becker, 1999). Residents of the community participated in a scientific effort to measure the amount of water captured by the forest and then percolated into their own underground water supplies. The community and the local NGO also prepared a video about their local forest that enabled most members of the community to come to a different understanding of the value of the forest, the danger of its overharvesting, and the benefits they would achieve by finding an effective way of preserving part of their forest for the future. With this kind of facilitative external assistance, the common understanding of benefits and costs changed in the community, and they constructed rules to regulate the use of their forests to achieve a more sustainable pattern.

This case demonstrated that there is no fixed relationship between the size, location, and shape of a forest and the perceptions that individuals hold about these variables. The relationship between perceptions and reality is itself potentially alterable through collective action. But when a forest is located at a substantial distance, this physical factor increases the difficulty of achieving a common understanding of likely benefits and increases the cost of achieving successful local, collective action.

Becker and León's analysis (chapter 7) shows quite clearly that the attributes of the Yuracaré have, until recently, allowed them to create and maintain institutions that resulted in the successful management of their common forests. In fact, it could be argued that Yuracaré possessed all of the user attributes: they depended significantly on the forest, they had a common understanding about the forest (about how to use the forest, which plants attracted wild game, and what activities were needed to manage the forest), they expected to remain in the forest, they shared a similar distribution of interests, they trusted each other to a great extent, they constructed rules free from others' interference, and they had a long history of organization. These attributes helped to reduce the costs that they face in constructing and maintaining a set of institutions that have sustained themselves and their forest resources for centuries.

The recent emergence of a commercial timber industry near the Yuracaré has affected some of these attributes, which may also affect the group's ability to manage their forest resources well. The timber market allowed some individuals within the Yuracaré to gain more than others, affecting their common understanding, their distribution of interests, and their levels of mutual trust. With continued urban growth in the area, their discount rate about the forest may also erode over time. As the costs associated with these attributes rise, the ability of the Yuracaré to maintain their previously successful institutions will be challenged.

Varughese's study of 18 communities in Nepal (chapter 8) offers a direct test of whether one of the popular explanations of deforestation explains the difference in forest conditions in the rural areas of Nepal. An increasing population is one of those "obvious" explanations given for why many countries are facing massive deforestation in contemporary times. Since Varughese found that over 65 percent of the forests whose conditions are improving also have an above-average population growth and that 55 percent of the forests whose conditions are worsening have below-average population growth, he concludes that there is not a general relationship between population growth and forest conditions in these 18 communities (see also Fairhead and Leach, 1996, whose evidence also challenges this proposition). On the other hand, Varughese finds strong support for McKean's assessment that common-property institutions can frequently be more effective than other forms of property for common-pool resources, and particularly for forest resources. He finds a high level of association between the degree of collective activity existing in a community and the condition of their forests. Collective activity is manifested in innovative ways to cope with large user-group sizes, for instance. Creating subcommittees and subgroups to deal with coordination of large memberships was one way. And, in his larger study, Varughese (1999) also examines how heterogeneity affects the likelihood of collective action and better forest conditions. He examines the impact of disparities in wealth, in distance to the forest, and in the number of women in decision-making positions in forest groups and differences in ethnicity on the likelihood of higher levels of organized collective action. What he finds is most interesting. While there are examples of groups that are heterogeneous in regard to diverse characteristics that are highly successful in their local organization, others are only moderately successful, and still others have failed to gain any effective organization. In other words, there is no obvious relationship between heterogeneity and successful organization.

Varughese also finds that the more heterogeneous groups that have organized themselves have developed various kinds of ingenious institutional arrangements for the purpose of reducing the potential divisiveness that comes from heterogeneities. Thus, several of the more heterogeneous and successful groups have created several types of membership so that those with diverse interests could participate in different ways. In one group, for example, the owners of tea shops have a large demand for fuelwood but little time to participate in the monitoring or maintenance of the forest. This group created a special membership category whereby those who could not participate pay more for membership and pay for the wood they obtain while not having to participate heavily in forest activities. The other category gains the benefit of funds that can be used for a variety of community purposes including additional forest monitors and training sessions. Thus, having sufficient autonomy to develop their own rules and experiment with them over time is indeed an important attribute that successful user groups in Nepal and India have to some extent. As more and more is learned about these groups, and why some are more successful than others, it will be possible to undertake still more systematic research related to the relative importance of the attributes discussed above as they affect the perceived benefits and costs of collective action. It will also be possible to better inform policymakers about institutional support structures that would facilitate participatory forest management.

Looking Ahead

As we discuss in the introduction, this book is an initial progress report for the IFRI research program. The long-term goal of this program is an analysis based on a large number of cases and repeated visits to the same locations. As of this writing, our sample size has increased to 104 sites including 173 forests, 226 settlements, and 3,780 forest plots (in which over 62,000 trees have been identified and measured). We have been able to make repeat visits to six sites in Nepal and about the same number in Uganda. These cases will give us a better opportunity to examine the importance of the attributes of the forests and the users as we all strive for better explanations of the emergence and performance of forest institutions at the local level. While many other papers drawing on the IFRI research program are in progress, so far none of them negate any of the findings reported in this volume.

IFRI researchers have also already begun to link local-level analyses with higher-level phenomena. Clearly, national and regional governments affect the institutions that govern forests at the local level, and it is equally clear that larger-scale forests require different kinds of institutions for their governance—including some forms of comanagement involving both local users and governmental officials in their governance. IFRI researchers are now employing tools in addition to the protocols for this work, including remotely sensed images and analyses based on geographic information systems and the modeling of agent-based behavior. Association with the Center for the Study of Institutions, Population, and Environmental Change has made it possible to expand our number of sites in the Western Hemisphere and to link this work to the larger research community interested in land use and land-cover change and in global environmental change.

Consequently, we end this book with a promise to continue and expand the research program reported on herein, the effort to link the empirical results to evolving theories about human organization at multiple levels of analysis, and to continue relating analyses of human behavior with their impact on forest ecosystems.

References

Agrawal, Arun. 1994. "Rules, Rule Making, and Rule Breaking: Examining the Fit between Rule Systems and Resource Use." In *Rules, Games, and Common-Pool Resources,* ed. Elinor Ostrom, Roy Gardner, and James Walker, 267–82. Ann Arbor: University of Michigan Press.

Baland, J. M., and J. P. Platteau. 1996. *Halting Degradation of Natural Resources: Is There a Role for Rural Communities?* Oxford: Clarendon Press.

Becker, Constance D. 1999. "Protecting a *Garúa* Forest in Ecuador: The Role of Institutions and Ecosystem Valuation." *Ambio* 28(2) (March): 156–61.

Bergstrom, T., L. Blume, and H. Varian. 1986. "On the Private Provision of Public Goods." *Journal of Public Economics* 29: 25–49.

Blair, Harry. 1996. "Democracy, Equity, and Common Property Resource Management in the Indian Subcontinent." *Development and Change* 27(3): 475–99.

Buchanan, James M., and Gordon Tullock. 1962. *The Calculus of Consent.* Ann Arbor: University of Michigan Press.

Cernea, Michael. 1989. "User Groups as Producers in Participatory Afforestation Strategies." World Bank Discussion Papers No. 70. Washington, DC: World Bank.

Chan, K. S., S. Mestelman, R. Moir, and R. A. Muller. 1996. "The Voluntary Provision of Public Goods under Varying Income Distributions." *Canadian Journal of Economics* 19: 54–69.

Dayton-Johnson, J., and P. Bardhan. 1998. "Inequality and Conservation on the Local Commons: A Theoretical Exercise." Working paper. Department of Economics, University of California, Berkeley.

Fairhead, James, and Melissa Leach. 1996. *Misreading the African Landscape: Society and Ecology in a Forest-Savanna Mosaic.* Cambridge: Cambridge University Press.

Hardin, Russell. 1982. *Collective Action.* Baltimore, MD: Johns Hopkins University Press.

Lam, Wai Fung. 1998. *Governing Irrigation Systems in Nepal: Institutions, Infrastructure, and Collective Action.* Oakland, CA: ICS Press.

Molinas, José R. l998. "The Impact of Inequality, Gender, External Assistance and Social Capital on Local-Level Collective Action." *World Development* 26(3): 413–31.

Olson, Mancur. 1965. *The Logic of Collective Action: Public Goods and the Theory of Groups.* Cambridge: Harvard University Press.

Ostrom, Elinor. 1990. *Governing the Commons: The Evolution of Institutions for Collective Action.* New York: Cambridge University Press.

———. 1992a. *Crafting Institutions for Self-Governing Irrigation Systems.* San Francisco: ICS Press.

———. 1992b. "The Rudiments of a Theory of the Origins, Survival, and Performance of Common-Property Institutions." In *Making the Commons Work: Theory, Practice, and Policy,* ed. Daniel W. Bromley et al., 293–318. San Francisco: ICS Press.

———. Forthcoming. "Reformulating the Commons." In *The Commons Revisited: An Americas Perspective,* ed. Joanna Burger, Richard Norgaard, Elinor Ostrom, David Policansky, and Bernard Goldstein. Washington, DC: Island Press.

Ostrom, Elinor, Roy Gardner, and James Walker. 1994. *Rules, Games, and Common-Pool Resources.* Ann Arbor: University of Michigan Press.

Schlager, Edella. 1990. "Model Specification and Policy Analysis: The Governance of Coastal Fisheries." Ph.D. diss., Indiana University, Bloomington.

Silva, Eduardo. 1994. "Thinking Politically about Sustainable Development in the Tropical Forests of Latin America." *Development and Change* 25(4): 697–721.

———. 1997. "The Politics of Sustainable Development: Native Forest Policy in Chile, Venezuela, Costa Rica and Mexico." *Journal of Latin American Studies* 29: 457–93.

Tang, Shui Yan. 1992. *Institutions and Collective Action: Self-Governance in Irrigation.* San Francisco: ICS Press.

Varughese, George. 1999. "Villagers, Bureaucrats, and Forests in Nepal: Designing Governance for a Complex Resource." Ph.D. diss., Indiana University, Bloomington.

Wade, Robert. 1994. *Village Republics: Economic Conditions for Collective Action in South India.* San Francisco: ICS Press.

World Bank. 1991. *The Forest Sector.* World Bank Policy Paper. Washington, DC: World Bank.

Appendix

International Forestry Resources and Institutions Research Strategy

Elinor Ostrom and Mary Beth Wertime

The International Forestry Resources and Institutions (IFRI) research program is a long-term effort to establish an international network of collaborating research centers (CRCs) that will

• Continuously monitor and report on forest conditions, plant biodiversity, and rates of deforestation in a sample of forests in their country or region;

• Continuously monitor and report on the activities and outcomes achieved by community organizations; local, regional, and national governments; businesses; nongovernmental organizations (NGOs); and donor-managed projects in their country or region;

• Analyze how socioeconomic, demographic, political, and legal factors affect the sustainability of ecological systems;

• Prepare policy reports of immediate relevance for forest users, government officials, NGOs, donors, and policy analysts;

• Build substantial in-country capacity to conduct rigorous and policy-relevant research relying on interdisciplinary teams already trained in advanced social and biological scientific methods; and

• Prepare training materials that synthesize findings for use by officials, NGOs, forest users, and students.

This Research Strategy was originally drafted in the initial planning phases of the project in 1994. It is appended to this volume of papers from the IFRI research program so that readers can understand the design of the overall program as well as the findings from some of the initial studies.

The Problem

Drastic measures to halt the alarming rates of deforestation, especially in the tropical forests of Central and South America, Asia, and Africa, are regularly proposed by officials, scholars, and those concerned with environmental issues. The term *crisis* often appears in the titles of scientific reports.[1] Noted scholars speak about "catastrophes about to happen"[2] or "mass extinction episodes" (Myers, 1988, 28). Indeed, projected rates of population growth, deforestation, and species loss are startling:

• The world's human population is predicted to be 10 billion by the year 2025 and 14 billion by the year 2100.[3]

• Most tropical forests "will be entirely lost or reduced to small fragments by early in the next century" (Task Force on Global Biodiversity, 1989, 3). "[P]rimary ancient forest areas are being destroyed at accelerating rates. At best, they are replaced by secondary forests which offer impoverished biodiversity, and, at worst, they are taken over by desertification" (Chichilnisky, 1994, 4).

• One-quarter to one-half of the earth's species will become extinct by 2020.[4]

These losses are often attributed to a set of causes that appear to vary depending on institutional affiliation, academic persuasion, or business or economic concern. Many individuals and environmental groups view commercial logging as the cause of deforestation (Task Force on Global Biodiversity, 1989, 3).[5] Shifting or new cultivation is viewed as the primary cause by scholars in other narratives.[6] Excessive energy consumption is cited by others. Population increase is considered by many to be a prime candidate causing deforestation and other environmental harms.[7]

A singular view of the cause is frequently paired with a singular view of the solution. Preservationists have often addressed the problem through "save and preserve" solutions. [Maintaining the position that strict actions must be taken to preserve the old-growth forests and the diversity of plant and animal life, proponents of this argument push for protected areas where certain activities, such as logging, are prohibited and species such as the spotted owl are protected.]

Policy analysts often recommend changes in international agreements or shifts in national policy as a solution. At the United Nations Conference on Environment and Development (UNCED) held in Rio de Janeiro in June of 1992, three major policy documents were produced at the conference (the Rio Declaration, Agenda 21, and the Forest Principles), and two conventions were released for signature (the Convention on Biological Diversity and the Convention on Climate Change). All of these documents proposed the adoption of international standards to regulate the use and management of natural resources—particularly forest resources, so as to enhance their diversity and sustainability over time.[8]

National governments have adopted government and industry reforestation schemes, forest-based industrial developments, and forestry action plans. National policies include changing forest commons into private land, assigning governments the responsibility of managing reserves and severely limiting access to these reserves, and prescribing community nurseries of predetermined tree species in rapidly changing environments—without regard for indigenous people, their changing environments, and methods of management of forest resources.

Agreement seems to exist about the need for immediate action. Less agreement exists about which policies will lead to actual improvements. A common theme in the evaluations of national and international efforts to stem the rates of deforestation is that many of these programs actually "accelerate the very damage their proponents intend to reverse" (Korten, 1993, 8).[9]

If the programs that are supposed to stem deforestation tend to accelerate it, something is wrong. The IFRI research program will attempt to ascertain what is wrong and provide better answers to the question of how to reduce deforestation and loss of biodiversity in many different parts of the world. In our efforts to understand what is wrong, we have identified three problems: (1) knowledge gaps, (2) information gaps, and (3) the need for greater assessment capabilities located in countries with substantial forest resources.

• A *knowledge gap* is the lack of an accepted scientific understanding about which variables are the primary causes of deforestation and biodiversity losses and how these variables are linked to one another. Policies that suggest ways to improve the effects of deforestation are often based on a model or theory about why deforestation is accelerating. However, the current status of theoretical explanations of the causes of deforestation and biodiversity losses is in flux. No agreement exists within the scientific community concerning which of multiple contending models of deforestation and biodiversity loss are empirically valid.

• An *information gap* is a lack of reliable data about specific policy-relevant variables in a particular time and location. In other words, the data needed to test competing theories of deforestation and biodiversity losses are not generally available. Detailed data about forest conditions within a country that are important for policy making are also not available.

• *Assessment capability* is the presence of permanent in-country centers with interdisciplinary staffs trained in rigorous forest mensuration techniques, participatory appraisal methods, institutional analysis, statistics, qualitative analysis, geographic information systems (GIS), and database management.

Alternative Approaches to Solving the Problem

Within the United States, the Committee on Environment and Natural Resources (CENR) of the National Science and Technology Council focused on the need for a better scientific foundation for future policy initiatives. CENR held a National Forum on Environment and Natural Resource R&D at the National Academy of Sciences in late March of 1994 in Washington, D.C. The Forum brought together representatives from industry, academia, nongovernmental organizations, Congress, and state and local governments to articulate their views on the strategy and priorities for issues related to environmental change. The Forum reached several conclusions about critical research needs that are relevant to the design of the IFRI research program, including the following:

• An improved understanding of the environmental issues requires a long-term commitment to a balanced research program of systematic observations (monitoring), data and information systems, process studies, and predictions (CENR, 1994, 5).

• The areas most in need of augmentation are:

· the scientific basis for integrated ecosystem management,

· the socioeconomic dimensions of environmental change,

· science policy tools,

· observations, and information and data management, and

· environmental technologies (CENR, 1994, 5).

When focusing on the socioeconomic dimensions of environmental change, the Forum identified specific research that needed substantial augmentation and emphasis. These included efforts to

• Understand the societal drivers of environmental changes, including the analyses of the environmental impacts of various patterns and growth of population, economic growth, and international trade;

• Promote policy analysis, including the design, comparison, and ex post evaluation of the effectiveness of policy alternatives to prevent, ameliorate, or manage environmental problems;

• Promote the analysis of environmental goals, encompassing the concepts of distributive justice, procedural fairness, community participation, and economic well-being; and

• Promote the analysis of the barriers to the diffusion of environmentally beneficial technologies (CENR, 1994, 6).

These critical research needs are challenging and require diverse approaches. One approach is that of global monitoring, relying primarily on national inventories and satellite imagery. Major progress to implement this approach has been taken by the United Nations Food and Agriculture Organization (1993). A second approach is to link permanent forestry and agroforestry research stations to foster more rapid exchange of scientific findings about how ecological systems are affected by (and affect) climate changes, increased pollution levels, and other environmental threats. Efforts of the U.S. National Science Foundation (NSF) to create such linkages have been successfully initiated.

A third approach—the one taken by the IFRI research program—complements the first two approaches and generates policy-relevant information not available from other strategies. The IFRI program provides an interdisciplinary set of variables about forest management and use that is collected near the forest in relationship to the local communities utilizing and governing the forest. The effects of district, national, and international policies as they impact on a local setting can be assessed through this effort. The results of IFRI studies provide in-country information for policymakers at the local, district, regional, and national levels. This information will be collected by researchers who are deeply familiar with the local settings rather than collected from secondary sources that are compiled by international organizations or by national agencies drawing on various sources of externally compiled information. The IFRI research program relies on the building of a permanent international network of CRCs. Each CRC will

• Design a long-term monitoring plan to include a sample of forests located in different ecological zones, managed by diverse institutional arrangements, and located near centers of intense population growth as well as in more remote regions;

• Conduct rigorous evaluations of projects undertaken to reduce deforestation, increase local participation, encourage ecotourism, change forest-tenure policies, implement new taxes or incentives, or in some way attempt to improve the incentives of officials and citizens to enhance and sustain forest resources and biodiversity;

• Provide useful and rapid feedback to officials and citizens about conditions and processes in particular forests of relevance to them;

• Archive data about environmental and institutional variables in a carefully designed database to be used within each country and to be shared among the participating research centers;

• Conduct analyses of those policies and institutional arrangements that perform best in particular political-economic and ecological settings; and

• Prepare materials of relevance for in-service training as well as for educational curricula.

Goals and Outcomes: Addressing Knowledge and Information Gaps and Building Assessment Capacities

The goals of the IFRI program are to (1) address the issue of knowledge gaps by seeking ways to enhance interdisciplinary knowledge, (2) address information gaps by providing a means to ground-truth aerial data and spatially link forest use to deforestation and reforestation, and (3) address the need for greater assessment capabilities by building capacity to rigorously collect, store, analyze, and disseminate data in participating countries.

Goal: Addressing Knowledge Gaps

Any system of interaction involving a relatively large number of variables that relate to one another over time with complex feedback loops is immensely more difficult to understand and control than simple systems tackled in more mechanistic areas, such as in classical physics. Human uses of forest resources involve a large number of potentially relevant variables that operate over time with complex feedback loops. Effective policy interventions are elusive until an empirically warrantable consensus is attained about the set of important variables that impinge on deforestation and biodiversity losses.

Recent attempts to understand processes leading to general environmental harms involve multivariable models. Paul and Anne Ehrlich (1991, 7), for example, propose a three-variable causal model:

$$I = P \times A \times T,$$

where I is impact on the environment, P is population size, A is affluence (as measured by levels of consumption), and T is technologies employed.

An alternative model developed by Grant (1994) for UNICEF to capture processes occurring primarily in developing countries is the PPE spiral where poverty and population pressures are viewed as reinforcing one another and jointly impinging on environmental conditions, while all three factors—population, poverty, and environment—affect and are affected by political instability.

The extent of the knowledge gap becomes apparent on careful examination of these two recent and respected models. They disagree on the size of the relationship between poverty on environmental variables.[10] The Ehrlichs include population *size* in their model, which is a state variable operationalized by either population density or the total number of people. The UNICEF model identifies population *growth* rather than current size. Technology appears in the Ehrlich model but not in the

UNICEF model. Political instability appears in the UNICEF model but not in the Ehrlich model. The logical places to intervene are different depending on which model best describes the world. If one accepts the Ehrlichs' view, one should focus attention on the most affluent countries ignoring political instability. Accepting the UNICEF view, one would focus on the poorest countries and stress the impact of political instability.

The effect of opening a region to increased market pressures is also a matter of debate in the literature. Many scholars presume that integrating local resource systems into larger markets by building roads and market centers increases the temptation that local users face to overharvest (see, for example, Agrawal, 1994). On the other hand, William Ascher (1995) argues that providing the poor in remote regions with better access to income-earning activities reduces their need to overuse forest resources and encourages a longer time horizon in making decisions about the use of local resources (see also Fox, 1993).

The knowledge gap is illuminated further by an important study by Robert T. Deacon (1994) on "Deforestation and the Rule of Law in a Cross-Section of Countries."[11] Using FAO estimates of forest cover in 1980 and 1985 to measure the proportionate rate of deforestation between 1980 and 1985, Deacon first examines the impact of population growth. He finds, in support of the UNICEF model, that a "one percent increase in population during 1975–1980 is associated with a proportionate forest cover reduction of 0.24–0.28 percent during 1980–85" (Deacon, 1994, 8). Supportive of the Ehrlich model, Deacon also finds that a "given rate of population growth is associated with a higher deforestation rate if it occurs in a high income country than in a low income country." While Deacon finds significant relationships, population change accounts for only a small proportion of the variance of deforestation (R^2 between .08 and .14).

The primary reason that Deacon undertakes this analysis, however, is to examine the impact of unstable or weakly enforced legal systems on deforestation. The decision to consume forest resources rapidly or to conserve them so as to yield a perpetual stream of future returns is an investment decision. Deacon (1994, 3) argues that investments will be made only when those who make a sacrifice not to harvest immediately are assured they will receive the future benefits of their actions: "When legal and political institutions are volatile or predatory, the assurance is lowered and the incentive to invest is diminished." Consequently, Deacon analyzes variables that reflect political instability and the presence of centralized national governments. These variables are positively associated with deforestation, and the proportion of the variance explained rises (R^2 between .19 and .21). Political and institutional variables account for as much or more variance in deforestation as population density. In the 120 countries included in his analysis, the size of the association between population growth and deforestation is reduced when political and institutional variables are included. The association falls substantially in low- and middle-income countries.[12] Deacon's analysis is pathbreaking because it is a rare effort to undertake a systematic analysis of the relative role of population density and institutional variables. He demonstrates that both have an impact on rates of deforestation. What his analysis also shows, however, is that factors affecting 80 percent of the variance in deforestation at a national level are not accounted for. This is a substantial knowledge gap.

While knowledge gaps about relationships at a national level remain immense, greater progress has been achieved in gaining a shared and empirically validated understanding of relationships at a more micro or subnational level. In the mid-1980s, the National Academy of Sciences (NAS) established a Panel on Common Property Resources. Since then, many theoretical and empirical studies of diverse institutional arrangements for governing and managing small- to medium-size natural resources have enabled scientists to achieve a growing consensus.[13] Scholars from diverse disciplines now tend to agree that the users of small- to medium-size natural resources are potentially capable of self-organizing to manage these resources effectively, whether jointly with national governments or with considerable autonomy. Researchers have even identified localities within countries where local users have organized themselves effectively enough that they have improved forest conditions when faced with increasing population density.[14]

There are several reasons why local users may more effectively manage resources than national agencies. One reason is the immense diversity of local environmental conditions that exist within most countries. The variation in rainfall, soil types, elevation, scale of resource systems, and plant and animal ecologies is large, even in small countries. Some resources are located near to urban populations or a major highway system, and others are remote. Given environmental variety, rule systems that effectively regulate access, use, and the allocation of benefits and costs in one setting are not likely to work well in radically different environmental conditions. Efforts to pass national legislation establishing a uniform set of rules for an entire country are likely to fail in many of the locations most at risk. Users managing their resources locally may be a more effective way of dealing with immense diversity from site to site.

A second reason for the potential advantage of local organization in coping with problems of deforestation and biodiversity losses is that the benefits local users may obtain from careful husbanding of their resources are potentially greater when future flows of benefits are appropriately taken into account. At the same time, the costs of monitoring and sanctioning rule infractions at a local level are relatively low. These advantages occur, however, only when local users have sufficient assurance that they will actually receive the long-term benefits of their own investments.

While there is agreement that the potential for effective organization at a local level to manage some of the small- to medium-size forests exists in all countries, local participants do not uniformly expend the effort needed to organize and manage local forests, even when given formal authority. Some potential organizations never form at all. Some do not survive more than a few months. Others organize but are not successful. Others are dominated by local elite who divert communal resources to achieve their own goals at the expense of others (Arora, 1994). In some cases, the natural forest must be almost completely gone before local remedial actions are taken. These actions may be too late. Still others do not possess adequate scientific knowledge to complement their own indigenous knowledge. Making investment decisions related to assets that mature over a long time horizon (25 to 75 years for many tree species) is a sophisticated task whether it is undertaken by barely literate farmers or Wall Street investors. In highly volatile worlds, some organize themselves more effectively and make better decisions than others.

Thus, the romantic view that anything local is better than anything organized at a national or global scale is not a useful foundation for a long-term effort to improve understanding of what factors enhance or detract from the capabilities of any institutional arrangement to govern and manage forest resources wisely. Any organization or group faces a puzzling set of problems when it tries to govern and manage complex multispecies (including *Homo sapiens*), multiproduct resource systems whose benefit streams mature at varying rates. Any organization or group will face a variety of environmental challenges stemming from too much or too little rainfall to drastic changes in factor prices, population density, or pollution levels. Consequently, essential knowledge can be gained from a carefully designed, systematic study of how many different types of institutional arrangements, including nascent groups, indigenous communal organizations, formal local governments, NGOs, specialized forest and park agencies, and national ministries, cope with diverse types of forest resources. Much is to be learned from both successes and failures. And, since we intend to use multiple performance measures, we expect to find some forest-governance and -management systems that are evaluated positively in regard to some evaluative criteria (such as the maintenance of forest density and species richness), but not necessarily in regard to others (such as gender representation, financial accountability, adaptability over time, or transparency of decision-making processes).

Outcome: Enhancing Interdisciplinary Knowledge Prior theoretical and empirical studies provide an initial set of hypotheses about general factors that we expect to find associated with the more successful forest-governance and -management systems (see Ostrom, 1990; McKean, 1992; Moorehead, 1994). Thus, the IFRI research program begins with an initial set of working hypotheses that will be revised, added to, and refined over time.

Our initial working hypotheses are that more effective organization to cope with the long-term sustainable management of forest resources will occur where

• Local forest users participate in and have continuing authority to design the institutions that govern the use of a forest system;

• The individuals most affected by the rules that govern the day-to-day use of a forest system are included in the group that can modify these rules;

• The institutions that govern a forest system minimize opportunities for free-riding, rent-seeking, asymmetric information, and corruption through effective procedures for monitoring the behavior of forest users and officials;[15]

• Forest users who violate rules governing the day-to-day uses of a forest system are likely to receive graduated sanctions from other users, from officials accountable to these users, or both;

• Rapid access is available to low-cost arenas to resolve conflict between users or between users and their officials;

• Monitoring, sanctioning, conflict resolution, and governance activities are organized in multiple layers of nested enterprises; and

• The institutions that govern a forest system have been stable for a long period and are known and understood by forest users.[16]

The variables in these hypotheses are all operationalized using multiple indicators in the IFRI research instruments. Further, we have included other variables noted in the literature as being of importance in explaining processes of deforestation and biodiversity loss. Additional variables are included in the design of this study based on the Institutional Analysis and Development framework,[17] which has served as the theoretical foundation for many of the successful prior studies of the governance and management of natural-resource systems undertaken by colleagues at Indiana University.

In the design of this study, we have also been concerned with how national and regional governments can enhance or detract from the capabilities of local entities by the kind of information they provide, by the assurance that they extend to ensure autonomy over the long run, by the provision of low-cost conflict resolution mechanisms, and by policies that allow localities to develop and keep financial resources that can be used to make local improvements. Detailed information about why some national policies tend to encourage successful self-organization and others discourage it will be provided. These results will help to reduce knowledge gaps about policy impacts and thus facilitate the development of more effective policies.

The IFRI research program is designed to examine relationships among the physical, biological, and cultural worlds in a particular location and the *de facto* rules that are used locally to determine access to and use of a forest. During data collection, researchers will use 10 research instruments. Examination of the physical world includes examination of the structure of forests and the species within. There are two research instruments that include rigorous forest mensuration methods to generate reliable and unbiased estimates of forest density, species diversity, and consumptive disturbances. Examination of cultural worlds includes gaining knowledge about patterns of socioeconomic and cultural homogeneity, number of individuals and groups involved, and diverse world views. Research conducted using a uniform set of variables using the best methods available for gaining reliable estimates of qualitative and quantitative data will enable scholars to analyze how different institutions work in the context of a large number of ecological, cultural, and political-economic settings. Diverse models of which variables and how they interact to affect behavior and outcomes will be posed, tested, and modified so that policies based on revised and tested models will have a higher probability of being successful than past efforts to reduce deforestation and stop biodiversity losses.

Goal: Addressing Information Gaps

Important steps have been taken in the last decade to increase the rigor and quantity of information known about forest cover and rates of deforestation and biodiversity losses in different parts of the world. In 1993, for example, the most "authoritative global tropical deforestation survey to be produced in more than a decade" (Aldhous, 1993, 1,390) was released by the United Nations Food and Agriculture Organization (1993). This FAO report attempts to document the extent of deforestation in tropical countries in an accurate fashion but repeatedly stresses the problems that the project staff faced in obtaining reliable information for the task. After examining the current state of information about forest conditions in tropical countries, the project found that

• There is considerable variation among regions with respect to completeness and quality of the information;

• There is considerable variation in the timeliness of the information: the data are about 10 years old, on average, which could be a potential source of bias in the assessment of change;

• Only a few countries have reliable estimates of actual plantations, harvests, and utilization although such estimates are essential for national forestry planning and policymaking;

• No country has carried out a national forest inventory containing information that can be used to generate reliable estimates of the total woody biomass volume and change;

• It is unlikely that the state and change information on forest cover and biomass could be made available on a statistically reliable basis at the regional or global level within the next 10 or 20 years unless a concerted effort is made to enhance the country capacity in forest inventory and monitoring (UNFAO, 1993, 5–6).

The report concludes its findings concerning information gaps by noting that "forest resource assessments are among the most neglected aspects of forest resource management, conservation and development in the tropics" (UNFAO, 1993, 6).

Outcome: Providing Key Ground-Truthed Information   The IFRI research program will immediately provide key information about variations in forest conditions and the incentives and behavior of forest users within countries participating in the IFRI network. This information is essential for policy analysis and to test theories addressing knowledge gaps. Focusing on a sample of forests located in diverse ecological regions and governed by different institutional arrangements greatly reduces the cost of monitoring as contrasted to national forest inventories. Further, it provides information about the variation of results achieved by different kinds of institutional arrangements.

Both quantitative and qualitative data will be collected about institutional arrangements, the incentives of different participants, their activities, and careful forest-mensuration techniques will be used to assess consequences in terms of density, species diversity, and species distribution. The general type of information to be collected at each site is listed in table A.1. This information will immediately be made available to forest users and government officials and used in regularized policy reports written by analysts who have a long-term stake in the success of the policies adopted. The results of projects adopted in one location can be compared with the results of other types of institutional arrangements in

Table A1
Data-collection forms and information collected

IFRI Form	Information Collected
Site Overview Form	Site overview map, local wage rates, local units of measurement, exchange rates, recent policy changes, interview information
Forest Form	Size, ownership, internal differentiation, products harvested, uses of products, master species list, changes in forest area, appraisal of forest condition
Forest Plot Form	Tree, shrub, and sapling size, density, and species type within 1-, 3-, and 10-meter circles for a random sample of plots in each forest, general indications regarding forest condition
Settlement Form	Sociodemographic information, relation to markets and administrative centers, geographic information about the settlement
User Group Form	Size, socioeconomic status, attributes of specific forest-user groups
Forest-User Group Relationship Form	Products harvested by user groups from specific forests and their uses
Forest Products Form	Details on three most important forest products (as defined by the user group), temporal harvesting patterns, alternative sources and substitutes, harvesting tools and techniques, harvesting rules
Forest Association Form	Institutional information about forest association (if one exists at the site), including association's activities, rules structure, membership, record keeping
Non-Harvesting Form	Information about organizations that make rules regarding a forest(s) but do not use the forest itself, including structure, personnel, resource mobilization, and record keeping
Organizational Inventory and Interorganizational Arrangements Form	Information about all organizations (harvesting or not) that relate to a forest, including harvest and governance activities

similar ecological zones within the same macropolitical regime. The data will also be archived in an IFRI-designed relational database so that changes in institutions, policies, activities, and outcomes can be monitored over time and across regions within one or more than one country. Data will be collected, owned, assessed, stored, and analyzed by each country's researchers. The IFRI research program fosters in-country development of information rather than sole reliance on the purchase of secondary data from international organizations. The program also encourages the development of "state-of-the-art" research conducted by researchers who have permanent roots in a country rather than those who come in from the outside.

Goal: Building Capacity for Assessment

The third major goal of the IFRI research program is to build in-country capacities to conduct forest and institutional assessments on a continuing basis. As the FAO (UNFAO, 1993) *Forest Resources Assessment* report cited above indicates, developing sustained efforts to gain an accurate picture of forest conditions or to build a valid understanding of what factors affect forest conditions is impossible *without building in-country assessment capabilities.* There are extraordinary researchers in each country with substantial capabilities that could be utilized in a sustained assessment program. These scholars may be located in different research institutions and separated by disciplinary barriers. Recent developments in the use of computers may not have been made available. For whatever reason, few countries have brought together interdisciplinary teams with extensive training in biology, environmental science, social sciences, and the use of computers to conduct regular assessments that can be used to fill information gaps and gain more valid understanding of the variables that affect rates of deforestation and loss of biodiversity.

The IFRI research program will work with a growing group of in-country research centers who obtain funding from donors and their own institutions to build their capabilities to become a permanent assessment center.

Outcome: Legacy of Long-Term Assessment Capabilities   In addition to addressing the problems of reducing knowledge and information gaps to enhance future forestry policymaking, the IFRI research program will leave a legacy in each participating country of a core research team that is well-trained in social and biological research methods and the computers to do analysis and manage complex forestry data sets.

Operational Methods of IFRI Research Program

As a research program, we envision a process of policy-relevant theoretical development, data collection, analysis, policy reporting, and training that is ongoing for the next decade or more. The overarching plan for the IFRI program is that future research goals and objectives will be addressed by a network of collaborating research centers (CRCs) and individual scholars who design and conduct studies within different countries in collaboration with colleagues at the Workshop in Political Theory and Policy Analysis and the Center for the Study of Institutions, Population, and Environmental Change (CIPEC). An IFRI CRC could be a research group associated with a university, a private association, a government research laboratory, or a consortium of individuals and agencies that have agreed to work together to collect, analyze, and archive IFRI data in a particular country or specific region of the world. Individual researchers who are working at a university or research institution completing their doctoral research or working independently may also be associated with IFRI.

The IFRI program includes a training model for each CRC that is intensive in the first two years. Each CRC will send key research personnel for a one-semester training program conducted by staff at Indiana University. This will be followed up with an in-country training program of a month's duration where the initial core set of researchers from a particular country or region are provided classroom and experiential training opportunities by Indiana University staff and by the local researchers who have just completed the semester in Bloomington. Pilot studies will be conducted soon after this initial training program has been completed. During the pilot studies, the Bloomington staff will be prepared to respond to methodological queries as in-country researchers discover the many complex and unexpected relationships using the methods they have just learned. As local staff become experts in the field administration, analysis, and archival of the data, further training will be taken over by those heading each of the CRCs. We also see a role for staff from one CRC visiting and working with staff from a second CRC so that the reliability of field methods and interpretations is enhanced.

Criteria for selecting CRCs will be based primarily on level of interest in solving forest resource problems from the bottom up, previous work on forest issues, and capacity to use the database system in an environment that enables communication between nongovernmental policymakers, forest users, governmental policymakers, scholars, and grant-writing capabilities. Demonstrated commitment to continuing, long-term research efforts will also be a criteria for CRC selection.

We envision that each CRC will go through several phases of relating to the Workshop/CIPEC and to other CRCs in the IFRI network. During the first phase—normally about a year in duration—one or two researchers, who will take a major role in the development of the CRC, would spend at least one semester at Indiana University. They will participate in a general course of study that includes both the underlying theoretical foundations for the IFRI research program and a specific training program on forest mensuration, PRA methods, detailed review of all IFRI research instruments, and joint fieldwork in a site near to Indiana University.

Ideally, during the summer following the above training program, researchers from the CRC and Indiana University will jointly train a larger group of researchers in data collection and entry methods and jointly conduct one to four pilot studies together. By working side by side in the conduct of the initial pilot studies, many of the problems that have faced earlier efforts to undertake multinational research efforts should be reduced. A key problem facing all such studies is how to establish and keep consistent data-collection methods so that the data placed in the same fields in the database are actually comparable. No amount of classroom instruction can cope effectively with this problem. Working side by side in the initial studies in each country is one method of substantially increasing the reliability and validity of the data collection efforts. Further, working out data-entry procedures and queries is equally important in developing a database that is robust and can be used over many years and by many participants.

After completing its first round of pilot studies, a new CRC will participate in a meeting of all CRCs. The first such meeting took place at Oxford University in mid-December of 1994, the second at Berkeley in June of 1996, and the third at CIFOR in November of 1997.

During initial training and pilot studies, the person taking primary responsibility for the development of a CRC in a particular country or region will begin work, in consultation with his or her own colleagues and with colleagues at Indiana University, on a research design for a continuing assessment program using the IFRI research instruments. Each monitoring plan will identify major knowledge and information gaps that will be addressed if the program outlined in it were undertaken. Where there are specific questions of importance in a particular country or region not covered by the IFRI research instruments, these will be supplemented with new instruments designed by the CRC and shared with other members of the network. The monitoring plan will be circulated among members of the IFRI network, to public officials and NGOs in the host country, and eventually to potential donors for funding. Once funding is received and the appropriate staff has been hired, the CRC will begin its own research program. Researchers from each CRC will visit other CRCs and undertake joint fieldwork with the researchers from other CRCs. This is another way that consistent data collection and interpretation can be undertaken in a multinational study.

Dissemination of Results

The results of the IFRI research program will be disseminated in multiple ways that include

• Immediate feedback of a site report to forest users and government officials interested in each site. The site report will contain a list of all plant species located in the forest(s) in the site, their relative importance and density, a history of each settlement, and an overview of the activities of user groups.

• Policy analysis reports issued by each CRC annually, summarizing the findings from the sample of forests and forest institutions included in that year's study. In the early years, these will be based on cross-sectional information. In the later years, these will contain analyses of developments over time. These reports will be widely circulated to policymakers, forest users, and scholars within each country and to all of the other IFRI CRCs.

• Special project reports comparing the activities and results obtained by a particular government, donor, or NGO-sponsored project with other institutional arrangements existing in similar ecological zones. These reports will also be widely circulated to policymakers, forest users, and scholars within each country and to all of the other IFRI CRCs.

• M.A. and Ph.D. theses completed by students who work at those CRCs that are located within universities or other in-country (or U.S.) universities. These studies will address some of the more difficult knowledge gaps that cannot be addressed in the initial policy reports.

• Methodological reports written by CRC and Indiana University scholars addressing some of the difficult measurement problems involved in the conduct of a multicountry, over-time study of institutional, behavioral, as well as forest-condition variables. These will be circulated to interested researchers throughout the world.

• Scholarly publications submitted by CRC and Indiana University scholars to academic journals and university presses so that the findings become part of the generally available knowledge base for social scientists, foresters, biologists, and public-policy scholars.

• Synopses of policy reports and more analytical reports that will be made available through the Internet to a wide diversity of interested colleagues who are connected electronically.

• Training programs for public officials held at CRCs once the in-country database is sufficient to provide better evidence for in-country forest planning.

• Curricular materials prepared for introduction into undergraduate and graduate instruction in relevant disciplinary courses.

Initial reaction of forest users and government officials to IFRI research reports has been enthusiastic. The volume, of which this appendix is a part, is also an effort to make the results known to public officials, forest users, and scholars throughout the world. Members of the IFRI teams involved will be glad to hear from readers and learn what has, or has not, been useful in our initial series of studies.

Notes

1. For example, see Wilson (1985), Task Force on Global Biodiversity, Committee on International Science (1989).

2. Bruce Cabarle, manager of the Latin America Forestry Program at the World Resources Institute, recently commented: "There really is a catastrophe waiting to happen, both for the forests and the people who live off them" (in Alper, 1993).

3. United Nations Population Fund (1989) projections based on current levels of birth control use. The estimated population in the World Bank's World Development Report (1993: 268–69) for 2025 is, however, a more modest 8.3 billion. It is not unusual to find discrepancies this large in projected population figures given different assumptions about initial starting conditions and rates of change.

4. See Lovejoy (1980), Ehrlich and Ehrlich (1981), and Norton (1986). Reid and Miller (1989, 37–38) estimate that between 1990 and 2020, between 5 to 15 percent of all species would be lost.

5. See also discussion in Ascher (1993).

6. "It is this broad-scale clearing and degradation of forest habitats [by communities of small-scale cultivators] that is far and away the main cause of species extinctions" (Myers, 1988, 29).

7. For very recent views stressing the primary and simple role of population increases, see Rowe, Sharma, and Browder (1992, 39–40), Abernathy (1993), Fischer (1993), Holdren (1992), Ness, Drake, and Brechin (1993), and Pimental et al. (1994).

8. The "Houston Communique" issued in 1990 is also relevant. See description in Sedjo (1992, 16).

9. Korten is summarizing her evaluation of the impact of a "showcase loan" by the Asian Development Bank to support the reforestation of 358,000 hectares of land in the Philippines. Similar evaluations have been made of many national and international efforts (see, for example, Arnold and Stewart, 1989; Sen and Das, 1987; Apichatvullop, 1993; Shanks, 1990; Chambers, 1994; McNeely, 1988; Repetto, 1988; Repetto and Gillis, 1988).

10. This may be due to the fact that UNICEF focuses primarily on the developing world. But is the Ehrlich model limited primarily to the industrialized world?

11. Deacon did not set out to test either of the models proposed by the Ehrlichs or by UNICEF and made no reference to either of them. Deacon (1994, 2) stresses that the "causes of deforestation are not well understood" and that the causes posited by some analysts are absent in the discussions of others. Deacon's own view is that the insecurity of property rights is a major contributing factor to deforestation.

12. In low- and middle-income countries, a 1 percent increase in population during 1975 to 1980 is associated with a proportional forest cover reduction of 0.07 to 0.13 percent during 1980 to 1985.

13. Among the books that have been written since the NAS report that provide a foundation for this growing consensus are McCay and Acheson (1987), Fortmann and Bruce (1988), Wade (1994), Berkes (1989), Pinkerton (1989), Sengupta (1991), Ostrom, Feeny, and Picht (1993), Netting (1993), Ostrom (1990, 1992), Ostrom, Gardner, and Walker (1994), Blomquist (1992), Tang, (1991), and Thomson (1992).

14. These include the work of Fairhead and Leach (1992) in Guinée; Agrawal (1994) in India; Tiffen, Mortimore, and Gichuki (1994) in Kenya; Fox (1993) in Nepal; and Meihe (1990) in Senegal.

15. Free-riding behavior occurs when individuals do not contribute to the provision or production of a joint benefit in the hopes that others will bear the cost of participating and that the free-riders will receive the benefits without paying the costs. Rent-seeking occurs when individuals obtain entitlements that enable them to receive returns that exceed the returns they would receive in an open, competitive environment. Asymmetric information occurs when some individuals obtain information of strategic value that is not available to others. Corruption occurs when individuals in official positions receive personal side-payments in return for the exercise of their discretion.

16. These hypotheses are obviously stated in a very general manner. We are presently developing a working paper that specifies how more specific versions of these hypotheses could eventually be analyzed using the IFRI database.

17. See Kiser and Ostrom (1982), Oakerson (1992), Ostrom (1986), and Ostrom, Gardner, and Walker (1994).

References

Abernathy, V. 1993. *Population Politics: The Choices That Shape Our Future.* New York: Plenum Press.

Agrawal, Arun. 1994. "The Illogic of Arithmetic in Resource Management: Overpopulation, Markets and Institutions as Explanations of Forest Use in the Indian Himalayas." Paper presented at the conference Workshop on the Workshop, Workshop in Political Theory and Policy Analysis, Indiana University, Bloomington, June 16.

Aldhous, Peter. 1993. "Tropical Deforestation: Not Just a Problem in Amazonia." *Science* 259 (March 5): 1,390.

Alper, Joe. 1993. "How to Make the Forests of the World Pay Their Way." *Science* 260 (June 25): 1,895–896.

Apichatvullop, Yaowalak. 1993. "Local Participation in Social Forestry." *Regional Development Dialogue* 14(1) (Spring): 32–45.

Arnold, J. E. M., and W. C. Stewart 1989. *Common Property Resources Management in India.* Oxford: Oxford University, Oxford Forestry Institute.

Arora, Dolly. 1994. "From State Regulation to People's Participation: Case of Forest Management in India." *Economic and Political Weekly* (March): 691–98.

Ascher, William. 1993. "Political Economy and Problematic Forestry Policies in Indonesia: Obstacles for Incorporating Sound Economics and Science." Durham, NC: Center for Tropical Conservation, Report.

———. 1995. *Communities and Sustainable Forestry in Developing Countries.* San Francisco: ICS Press.

Berkes, Fikret, ed. 1989. *Common Property Resources: Ecology and Community-Based Sustainable Development.* London: Belhaven Press.

Blomquist, William. 1992. *Dividing the Waters: Governing Groundwater in Southern California.* San Francisco: ICS Press.

CENR (Committee on Environment and Natural Resources). 1994. "Draft R&D Strategy." Washington, DC: National Oceanic and Atmospheric Administration.

Chambers, Robert. 1994. "The Poor and the Environment: Whose Reality Counts." Paper prepared for the conference on Poverty Reduction and Development Cooperation held at the Centre for Development Research, Copenhagen, February 23–24.

Chichilnisky, Graciela. 1994. "Biodiversity and Economic Values." Summary of a paper presented at the conference Biological Diversity: Exploring the Complexities, University of Arizona, Tucson, March 25–27.

Deacon, Robert T. 1994. "Deforestation and the Rule of Law in a Cross-Section of Countries." Discussion Paper No. 94-23. Washington, DC: Resources for the Future.

Ehrlich, Paul R., and Anne H. Ehrlich. 1981. *Extinction: The Causes of the Disappearance of Species.* New York: Random House.

———. 1991. *Healing the Planet: Strategies for Resolving the Environmental Crisis.* Reading, MA: Addison Wesley.

Fairhead, James, and Melissa Leach, with Dominique Millimouno and Marie Kamano. 1992. "Forests of the Past? Archival, Oral and Demographic Evidence in Kissidougou Prefecture's Vegetation History." COLA Working Paper No. 1. Conakry, Guinea: Connaissance et Organisation Locales Agro-ecologiques.

Fischer, G. 1993. "The Population Explosion: Where Is It Leading?" *Population and Environment* 15(2): 139–53.

Fortmann, Louise, and John W. Bruce. 1988. *Whose Trees? Proprietary Dimensions of Forestry.* Boulder, CO: Westview Press.

Fox, Jefferson. 1993. "Forest Resources in a Nepali Village in 1980 and 1990: The Positive Influence of Population Growth." *Mountain Research and Development* 13(1): 89–98.

Grant, James P. 1994. *The State of the World's Children 1994.* Oxford: Oxford University Press for UNICEF.

Holdren, C. 1992. "Population Alarm." *Science* 255: 1,358.

Kiser, Larry L., and Elinor Ostrom. 1982. "The Three Worlds of Action: A Metatheoretical Synthesis of Institutional Approaches." In *Strategies of Political Inquiry,* ed. Elinor Ostrom, 179–222. Beverly Hills, CA: Sage.

Korten, Frances F. 1993. "The High Costs of Environmental Loans." *Asia Pacific Issues No. 7.* Hawaii: East-West Center.

Lovejoy, Thomas E. 1980. "A Projection of Species Extinctions." In *The Global 2000 Report to the President: Entering the Twenty-first Century,* G. O. Barney (Study Director), 328–31. Washington, DC: Council on Environmental Quality, U.S. Government Printing Office.

McCay, Bonnie J., and James M. Acheson. 1987. *The Question of the Commons: The Culture and Ecology of Communal Resources.* Tucson: University of Arizona Press.

McKean, Margaret A. 1992. "Success on the Commons: A Comparative Examination of Institutions for Common Property Resource Management." *Journal of Theoretical Politics* 4(3) (July): 247–82.

McNeely, Jeffrey A. 1988. *Economics and Biological Diversity: Developing and Using Economic Incentives to Conserve Biological Resources.* Gland, Switz.: IUCN.

Meihe, S. 1990. "Inventaire et Suivi de la Végétation dans les parcelles pastorales à 1990 et évaluation globale provisoire dl'essai de pâturage controlé après une période de 10 ans." Eshborn, Germany: GTZ, Research Report.

Moorehead, Richard. 1994. "Policy and Research into Natural Resource Management in Dryland Africa: Some Concepts and Approaches." International Institute for Environment and Development, London.

Myers, Norman. 1988. "Tropical Forests and Their Species: Going, Going . . ." In *Biodiversity,* ed. E. O. Wilson and Frances M. Peter, 28–35. Washington, DC: National Academy Press.

Ness, G., W. Drake, and S. Brechin. 1993. *Population-Environment Dynamics: Ideas and Observations.* Ann Arbor: University of Michigan Press.

Netting, Robert M. 1993. *Smallholders, Householders: Farm Families and the Ecology of Intensive, Sustainable Agriculture.* Stanford: Stanford University Press.

Norton, B. J., ed. 1986. *The Preservation of Species.* Princeton, NJ: Princeton University Press.

Oakerson, Ronald J. 1992. "Analyzing the Commons: A Framework." In *Making the Commons Work: Theory, Practice, and Policy,* ed. Daniel W. Bromley et al., 41–59. San Francisco: ICS Press.

Ostrom, Elinor. 1986. "An Agenda for the Study of Institutions." *Public Choice* 48: 3–25.

———. 1990. *Governing the Commons: The Evolution of Institutions for Collective Action.* New York: Cambridge University Press.

———. 1992. *Crafting Institutions for Self-Governing Irrigation Systems.* San Francisco: ICS Press.

Ostrom, Elinor, Roy Gardner, and James Walker. 1994. *Rules, Games, and Common-Pool Resources.* Ann Arbor: University of Michigan Press.

Ostrom, Vincent, David Feeny, and Hartmut Picht, eds. 1993. *Rethinking Institutional Analysis and Development: Issues, Alternatives, and Choices.* 2d ed. San Francisco: ICS Press.

Pimental, D., R. Harman, M. Pacenza, J. Pecarsky, and M. Pimental. 1994. "Natural Resources and an Optimal Human Population." *Population and Environment* 15(5): 347–69.

Pinkerton, Evelyn, ed. 1989. *Co-operative Management of Local Fisheries: New Directions for Improved Management and Community Development.* Vancouver: University of British Columbia Press.

Reid, Walter V., and Kenton R. Miller. 1989. *Keeping Options Alive: The Scientific Basis for Conserving Biodiversity.* Washington, DC: World Resources Institute.

Repetto, Robert. 1988. *The Forest for the Trees? Government Policies and the Misuse of Forest Resources.* Washington, DC: World Resources Institute.

Repetto, Robert, and Malcolm Gillis. 1988. *Public Policies and the Misuse of Forest Resources.* Washington, DC: World Resources Institute.

Rowe, R., N. Sharma, and J. Browder. 1992. "Deforestation: Problems, Causes, and Concerns." In *Managing the World's Forests,* ed. N. Sharma, 33–46. Dubuque, IA: Kendall/Hunt.

Sedjo, Roger A. 1992. "A Global Forestry Initiative." *Resources* no. 109 (Fall): 16–19.

Sen, D., and P. K. Das. 1987. "The Management of People's Participation in Community Forestry: Some Issues." London: ODI, Social Forestry Network Paper No. 4D, ODI, London.

Sengupta, Nirmal. 1991. *Managing Common Property: Irrigation in India and the Philippines.* London: Sage.

Shanks, E. 1990. "Communal Woodlots in Tanzania: Farmers' Response and an Evolving Extension Strategy." Social Forestry Network Paper No. 11C, ODI, London.

Tang, Shui Yan. 1991. "Institutional Arrangements and the Management of Common-Pool Resources." *Public Administration Review* 51(1) (January/February): 42–51.

Task Force on Global Biodiversity, Committee on International Science. 1989. *Loss of Biological Diversity: A Global Crisis Requiring International Solutions.* Washington, DC: National Science Board.

Thomson, James T. 1992. *A Framework for Analyzing Institutional Incentives in Community Forestry.* Rome: United Nations Food and Agriculture Organization.

Tiffen, Mary, Michael Mortimore, and F. N. Gichuki. 1994. *More People, Less Erosion: Environmental Recovery in Kenya.* New York: Wiley.

United Nations Food and Agriculture Organization (UNFAO). 1993. *Forest Resources Assessment 1990.* Rome: FAO.

United Nations Population Fund (UNFPA). 1989. *State of the World Population 1989.* New York: UNFPA.

Wade, Robert. 1994. *Village Republics: Economic Conditions for Collective Action in South India.* San Francisco: ICS Press.

Wilson, E. O. 1985. "The Biodiversity Crisis: A Challenge to Science." *Issues in Science and Technology* 2: 20–29.

World Bank. 1993. *World Development Report: Investing in Health.* New York: Oxford University Press.

Index